Table of Contents

iOS Sensor Apps with Arduino

Alasdair Allan

O'REILLY®

Beijing · Cambridge · Farnham · Köln · Sebastopol · Tokyo

iOS Sensor Apps with Arduino
by Alasdair Allan

Published by O'Reilly Media, Inc., 1005 Gravenstein Highway North, Sebastopol, CA 95472.

O'Reilly books may be purchased for educational, business, or sales promotional use. Online editions are also available for most titles (*http://my.safaribooksonline.com*). For more information, contact our corporate/institutional sales department: (800) 998-9938 or *corporate@oreilly.com*.

Editors: Shawn Wallace and Brian Jepson	**Cover Designer:** Karen Montgomery
Production Editor: Teresa Elsey	**Interior Designer:** David Futato

ISBN: 978-1-449-30848-3

[LSI]

1315842073

Preface

The iPhone (iPod touch and iPad) platform comes with a growing range of built-in sensors: GPS, accelerometer, magnetometer, and most recently a gyroscope. The devices also have near-ubiquitous data connections, whether via a local wireless hotspot or via carrier data, and user positioning via multiple methods, including the GPS. They make excellent hubs for a distributed sensor network.

However, until recently, it was actually quite difficult to interface these otherwise interesting devices and connect them to your iPhone. Apple's proprietary dock connector is a major stumbling block. This has now changed, and by the end of the book, you'll be able to use your phone as the hub of a sensor network, making it part of the Internet of Things.

Who Should Read This Book?

This book provides an introduction to connecting your iOS device to the real world. As part of that, we'll make extensive use of the Arduino (*http://arduino.cc/*) open source electronics prototyping platform. If you are a programmer who has had some experience with the iPhone before, this book will help you connect your iOS device to external hardware. If you are an experienced Mac programmer, already familiar with Objective-C, this book will give you an introduction to the hardware-specific parts of iPhone programming.

What Should You Already Know?

The book assumes some previous experience with the Objective-C language. Additionally, some familiarity with the iPhone platform is assumed. If you're new to the iPhone platform you may be interested in *Learning iPhone Programming* (O'Reilly), also by Alasdair Allan. Little or no familiarity with the Arduino platform is assumed or expected. This book is intended for Objective-C programmers who want to learn how to talk to external hardware. However, if you are totally unfamiliar with the Arduino platform, you might want to take a look at *Getting Started with Arduino* by Massimo Banzi (O'Reilly).

This book assumes a working knowledge of how to build and deploy applications onto your iPhone, iPod touch, or iPad. If you have no experience with iOS, you should probably read this book in conjunction with Learning iPhone Programming.

What Will You Learn?

This book will guide you through developing applications for the iOS platform that make use of the Redpark Serial Cable (*http://www.redpark.com/*) and the External Accessory framework to connect your iPhone, iPod touch, or iPad to any standard serial (RS-232) capable device. Beyond this, we'll also take a look at less official (but possibly more fun) ways to accomplish the same ends.

What's in This Book?

Chapter 1
> This chapter is intended for Objective-C programmers new to Arduino. It will introduce you to the platform and walk you through the hardware equivalent of "Hello World," the blinking LED. We'll also discuss how to use the serial connection between the Arduino and your development machine.

Chapter 2
> This chapter will introduce the Redpark Serial Cable and walk you through building two basic applications. The first will allow the Arduino to talk to the iPhone, and the second will allow the iPhone to talk to the Arduino.

Chapter 3
> This chapter walks through building a full-scale application for the iPad to allow arbitrary control over an Arduino from a user-friendly interface.

Chapter 4
> In this chapter, we will attach sensors to the Arduino and then use the serial cable to communicate the readings to the iPhone. We'll also look at the Core Plot library for graphing.

Chapter 5
> This chapter introduces the Digi XBee radio and walks you through connecting them to both the Arduino and iOS platforms.

Chapter 6
> This chapter looks at other ways to connect your iPhone or iPad to serial devices. We look at using the MIDI protocol, which is officially supported by the iOS SDK, but also at using the headphone jack of your iOS device as a soft modem.

Conventions Used in This Book

The following typographical conventions are used in this book:

Italic
> Indicates new terms, URLs, email addresses, filenames, and file extensions.

`Constant width`
> Used for program listings, as well as within paragraphs to refer to program elements such as variable or function names, databases, data types, environment variables, statements, and keywords.

`Constant width bold`
> Shows commands or other text that should be typed literally by the user.

`Constant width italic`
> Shows text that should be replaced with user-supplied values or by values determined by context.

 This icon signifies a tip, suggestion, or general note.

 This icon signifies a warning or caution.

Using Code Examples

This book is here to help you get your job done. In general, you may use the code in this book in your programs and documentation. You do not need to contact us for permission unless you're reproducing a significant portion of the code. For example, writing a program that uses several chunks of code from this book does not require permission. Selling or distributing a CD-ROM of examples from O'Reilly books does require permission. Answering a question by citing this book and quoting example code does not require permission. Incorporating a significant amount of example code from this book into your product's documentation does require permission.

We appreciate, but do not require, attribution. An attribution usually includes the title, author, publisher, and ISBN. For example: "*iOS Sensor Apps with Arduino*, by Alasdair Allan (O'Reilly). Copyright 2011 Alasdair Allan, 978-1-449-30848-3."

If you feel your use of code examples falls outside fair use or the permission given here, feel free to contact us at *permissions@oreilly.com*.

Almost all of the code in this book will not run on the iPhone or iPad Simulator and must be deployed and tested directly on your device.

Safari® Books Online

Safari Books Online is an on-demand digital library that lets you easily search over 7,500 technology and creative reference books and videos to find the answers you need quickly.

With a subscription, you can read any page and watch any video from our library online. Read books on your cell phone and mobile devices. Access new titles before they are available for print, and get exclusive access to manuscripts in development and post feedback for the authors. Copy and paste code samples, organize your favorites, download chapters, bookmark key sections, create notes, print out pages, and benefit from tons of other time-saving features.

O'Reilly Media has uploaded this book to the Safari Books Online service. To have full digital access to this book and others on similar topics from O'Reilly and other publishers, sign up for free at *http://my.safaribooksonline.com*.

How to Contact Us

Please address comments and questions concerning this book to the publisher:

O'Reilly Media, Inc.

1005 Gravenstein Highway North

Sebastopol, CA 95472

800-998-9938 (in the United States or Canada)

707-829-0515 (international or local)

707-829-0104 (fax)

We have a web page for this book, where we list errata, examples, and any additional information. You can access this page at:

http://oreilly.com/catalog/9781449308483

Supplementary materials are also available at:

http://www.programmingiphonesensors.com/

To comment or ask technical questions about this book, send email to:

bookquestions@oreilly.com

For more information about our books, conferences, Resource Centers, and the O'Reilly Network, see our website at:

http://www.oreilly.com

Acknowledgments

Everyone has one book in them. This is my third, or depending how you want to look at it, it isn't. This book, along with three other short ebooks on iOS and sensor technology that will be published over the next couple of months, will form the bulk of *iOS Sensor Programming (http://oreilly.com/catalog/0636920003250)*, which would probably be classed by most as my second real book for O'Reilly.

I'd like to thank my editor at O'Reilly, Brian Jepson, for holding my hand yet again. As ever, his hard work made my hard work much better than it otherwise would have been. I also very much want to thank my wife, Gemma Hobson, for her continued support and encouragement. Those small, and sometimes larger, sacrifices an author's spouse routinely has to make don't get any less inconvenient the second, or third, time around. I'm not sure why she let me write another; perhaps because I claimed to enjoy writing the first one so much. Thank you, Gemma. Finally to my son, Alex, as yet still too young to read what his daddy has written; hopefully this volume will keep you in books to chew on just a little longer.

CHAPTER 1

Introduction to the Arduino

Every so often a piece of technology can become a lever that lets people move the world, just a little bit. The Arduino is one of those levers. While it started off as a project to give artists access to embedded microprocessors for interactive design projects, I think it's going to end up in a museum as one of the building blocks of the modern world. It allows rapid, cheap prototyping for embedded systems. It turns what used to be fairly tough hardware problems into simpler software problems.

The Arduino, and the open hardware movement that has grown up with it and to a certain extent around it, is enabling a generation of high-tech tinkers to prototype new ideas with fairly minimal hardware knowledge.

The Arduino

While there are many other microcontroller platforms available, for a newcomer to physical computing, the simplest to work with is probably the Arduino (see Figure 1-1). This is in part due to the large community surrounding the platform, but also because the Arduino was designed from the ground up to be simple for newcomers.

The current revision of the Arduino board is known as the Arduino Uno. This board is based on the ATmega328 microcontroller. It has fourteen digital input/output pins, six of which can be used as Pulse Width Modulation (PWM) outputs, along with six more analog input pins. See Table 1-1.

Table 1-1. Technical specifications of the Arduino Uno board

Arduino Uno	
Microcontroller	ATmega328
Operating Voltage	5 V
Input Voltage (recommended)	7–12 V
Input Voltage (limits)	6–20 V
Digital I/O Pins	14 (6 provide PWM)

Figure 1-1. The Arduino Uno board with the ATmega328 microcontroller

Arduino Uno

Analog Input Pins	6
DC Current per I/O Pin	40 mA
DC Current for 3.3 V Pin	50 mA
Flash Memory	32 KB
SRAM	2 KB
EEPROM	1 KB
Clock Speed	16 MHz

The Arduino platform has everything needed to support the on-board microcontroller. It is attached directly to your Mac for programming using a USB connection, and it can be powered via the same USB connection or with an external power supply if you want to detach the board from your Mac once you've programmed it.

 To keep things short, this tutorial will assume you're using a recent Arduino board such as the Arduino Uno, Duemilanove, or Diecimila. However, if you're working with an older board, or one of the many Arduino-compatible clones, it's likely that little needs to be changed. See *http://arduino.cc/en/Main/Hardware* for links to the getting started guides for various other boards.

Powering the Board

The Arduino Uno can be powered via the USB connection or with an external power supply. Unlike with previous generations of the Arduino, the power source is selected automatically. If you're using an earlier model, you will have to manually change between USB and external power sources using a jumper on the board itself. This jumper is usually located between the USB and power jacks.

The board can operate on an external supply of 6 to 20 volts. The easiest option is to use a 9V battery or power adaptor, as these are commonly available.

Input and Output

Each of the 14 pins on the Uno can be used as an input or output. They operate at 5 V with a maximum current of 40 mA. Each pin also has an internal pull-up resistor of 20–50 kOhms, although this is disconnected by default.

Some pins have specialized functions. Perhaps the most important of these to us for this book are pins 0 and 1. These pins can be used to receive (RX) and transmit (TX) TTL serial data. These pins are connected to the corresponding pins of the ATmega8U2 and hence to the USB-Serial connection to your Mac.

Communicating with the Board

The ATmega328 provides UART TTL serial communication at 5 V, which is available on digital pins 0 (RX) and 1 (TX). The Arduino Uno has an ATmega8U2 chip on board that redirects this serial communication over USB, allowing the Arduino to appear as a virtual serial port to software on your Mac. If you're using an older board, such as the Duemilanove or Diecimila, these boards make use of the FTDI USB-to-serial driver chip to accomplish the same task, although unlike with the newer Uno board, you will be required to install a driver so that your Mac can see the board correctly.

Installing the Software

Download the latest version of the development environment from the Arduino.cc website (*http://www.arduino.cc/*).

The latest version of the Arduino development environment is available at *http://arduino.cc/en/Main/Software*. At the time of this writing, it was Arduino 022.

Like most Mac applications, the development environment comes as a disk image (.*dmg* file) that should mount automatically after you have downloaded it. If it doesn't, double-click on it to open it manually. After it is open, just click and drag the *Arduino.app* application into your */Applications* folder. If you're using an older board that requires that you install the FTDI USB-to-serial drivers, you should also double-click on the *FTDIUSBSerialDriver.mpkg* file included in the disk image to install the necessary drivers.

Once you've installed the development environment, eject the disk image by dragging it to the trash can and double-click on the *Arduino.app* application icon to start the IDE. You should see something that looks a lot like Figure 1-2.

Figure 1-2. The development environment

Connecting to the Board

Connect the Arduino to your Mac with an appropriate USB cable. In the case of an Arduino Uno, Duemilanove, or Diecimila, you'll need a USB-A (Mac) to USB-B (Arduino) cable, the same sort needed for most USB printers. The green power LED (labeled PWR) should go on, and if you're using the Arduino Uno, a dialog box will appear telling you that a new network interface has been detected. Just hit the Apply button. While the new interface will claim to be "Not Configured" if you inspect it in System Preferences, it is working correctly.

After connecting your board to your Mac, go to the Tools→Board menu item in the Arduino development environment and select your board from the list of supported boards in the drop-down menu (see Figure 1-3).

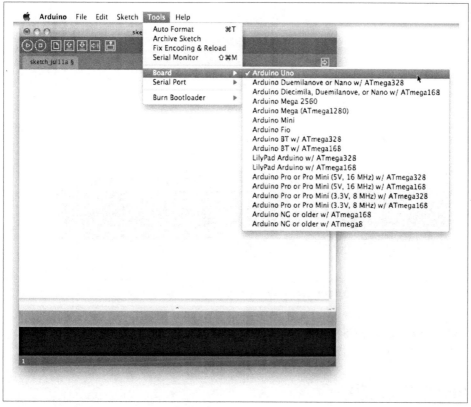

Figure 1-3. Selecting the correct board type

Then go to the Tools→Serial Port menu and select the correct serial port for your board (see Figure 1-4). On the Mac, the name will start with */dev/tty.usbmodem* for the Uno or */dev/tty.usbserial* for older boards.

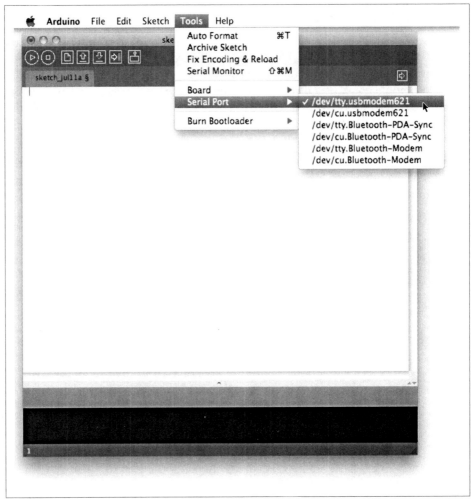

Figure 1-4. Choosing the correct serial port

If you're unsure which serial port corresponds to your board, you can always unplug and then replug the USB cable connecting your Mac to the Arduino board to see how the menu changes.

Blinking an LED

Now that we have our Arduino connected to our Mac, let's go ahead and walk through the hardware equivalent of "Hello World," the blinking LED.

An Arduino program is normally referred to as a *sketch*.

There are a number of example sketches included in the development environment. The one we're looking for can be found by selecting File→Examples→1.Basics→Blink. The example sketch will open in a new window; see Figure 1-5.

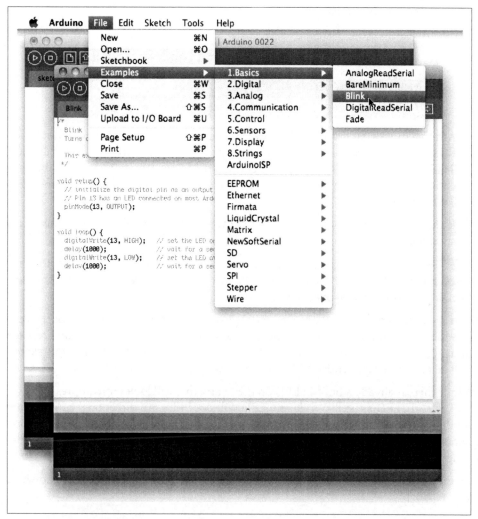

Figure 1-5. The Blink example

Every Arduino sketch consists of two parts: the setup and the loop. Every time the board is powered up or the board's reset button is pushed, the setup() routine is run. After that finishes, the board runs the loop routine. When that completes, and perhaps somewhat predictably, the loop() is run again and again. Effectively the contents of the loop() sit inside an infinite while loop.

Before we go ahead and build and deploy this example to our Arduino, let's take a look at the code:

```
void setup() {
  pinMode(13, OUTPUT);❶
}

void loop() {
  digitalWrite(13, HIGH);❷
  delay(1000);
  digitalWrite(13, LOW);❷
  delay(1000);
}
```

❶ Arduino pins default to INPUT. However, here we set pin 13 to behave as an OUTPUT pin. In this state, the pin can provide up to 40 mA of current to other devices. This is enough current to brightly light up an LED or run many sensors, but not enough current to run most relays, solenoids, or motors. See *http://arduino.cc/en/ Tutorial/DigitalPins*.

❷ If the pin has been configured for OUTPUT, its voltage will be set to the corresponding value: 5 V for HIGH, 0 V (ground) for LOW.

Effectively, then, this piece of code causes the voltage on pin 13 to be brought HIGH for a second (1,000 ms) and then LOW for a further second before the loop starts again, bringing the voltage HIGH once more.

As I mentioned before, some of the digital pins on the Arduino board are specialized; pin 13 is one of these. On most boards (including the Uno), it has an LED and resistor attached to it that's soldered onto the board itself. Therefore, the effect of this sketch will be to turn the embedded LED on and off with a periodicity of 1 second.

While there is already a built-in LED on the board, this will look more impressive if we add a "real" LED as well. While any LED will do, LEDs are directional components and must not be inserted backward. Look carefully at the two legs of the LED; one should be shorter than the other. The shorter leg corresponds to the ground, while the longer leg is the positive.

Insert the short leg into the GND pin, and the longer leg into pin 13, as shown in Figure 1-6.

Figure 1-6. An LED connected to pin 13; the short leg is inserted into the GND pin

Uploading the Sketch

The first step of getting the sketch ready for transfer to the Arduino is to click on the Verify/Compile button (see Figure 1-7). This will compile your code, checking it for errors, and then translate your program into something that is compatible with the Arduino architecture. After a few seconds, you should see the message "Done compiling" in the status bar and something along the lines of "Binary sketch size: 1018 bytes (of a 32256 byte maximum)" in the Notification area.

Once you see that message, go ahead and click on the Upload button (see Figure 1-7 again). This will initiate the transfer of the compiled code to the board via the USB connection.

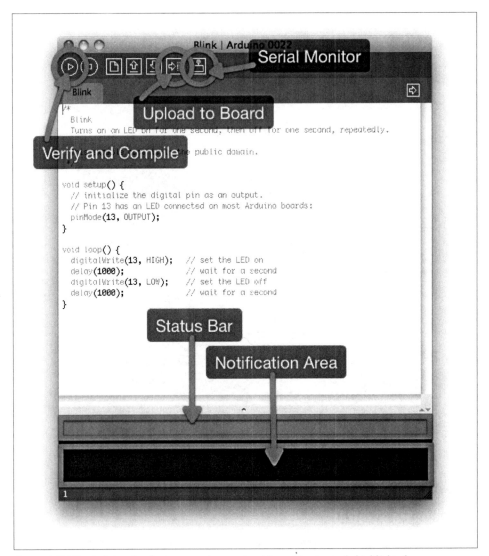

Figure 1-7. The Arduino development environment with various controls highlighted

Wait a few seconds; you should see the RX and TX LEDs on the board flashing as the data is transferred over the serial connection from your Mac to the board. If the upload is successful, the message "Done uploading" will appear in the status bar.

A few seconds after the upload finishes, you should see the pin 13 LED on the board (labeled L on the PCB) start to blink (in orange), along with the LED we inserted into pin 13. One second on, one second off. If you see that, congratulations, you've just successfully got the hardware equivalent to "Hello World" to compile and run on the Arduino board.

Making a Serial Connection

Now that we've seen how a basic Arduino sketch works, let's move on to sending and receiving data from the Arduino board. We'll need to know how to do this because this is how we'll control the Arduino or get sensor readings from it when we connect it to our iOS device. For now, however, we're going to use the Serial Monitor (see Figure 1-7 again), inside the development environment.

Open a new sketch by clicking File→New (⌘N) to open a new window:

```
void setup() {
  Serial.begin(9600);❶
}

void loop() {
  while (Serial.available() <= 0) {❷
    Serial.println("Hello world");❸
    delay(300);
  }
}
```

❶ Sets the data rate in bits per second (baud) for serial data transmission.

❷ Here we get the number of bytes available for reading from the serial port. If there are no bytes waiting in the buffer, we loop until some arrive.

❸ Finally, here inside the loop we print "Hello World" to the serial connection.

Effectively this code will send the string "Hello World" to the serial connection every 300 ms until the board receives a byte (character), at which point it will stop.

Save the contents of the sketch to a file using the File→Save (⌘S) menu item, and then click on the Verify button to compile your sketch. If all goes well, click the Upload button to upload it to the board. You should see the RX and TX LEDs light up as the code is transferred. When you see the "Done uploading" message in the status bar, click on the Serial Console button (see Figure 1-7 again) in the development environment to open the serial console.

Doing so will reset the Arduino, at which point you should see the phrase "Hello World" appear on the console every 300 ms (see Figure 1-8), accompanied by a flash of the TX LED on the Arduino board itself.

Entering a character or string in the text entry box and hitting the Send button will transmit those characters to the board, at which point the board should stop sending the string "Hello world" to the serial port.

Let's make that a bit more specific. Add the following highlighted lines to your sketch and re-upload it to the Arduino:

```
void setup() {
  Serial.begin(9600);
}
```

```
void loop() {
  while (Serial.available() <= 0) {
    Serial.println("Hello world");
    delay(300);
  }

  Serial.println("Goodbye world");
  while(1) { }
}
```

The serial console will close automatically when you upload new code to the board, so reopen it and you should see things much as before (Figure 1-8). However, this time if you send a string to the board you should receive the string "Goodbye world" back.

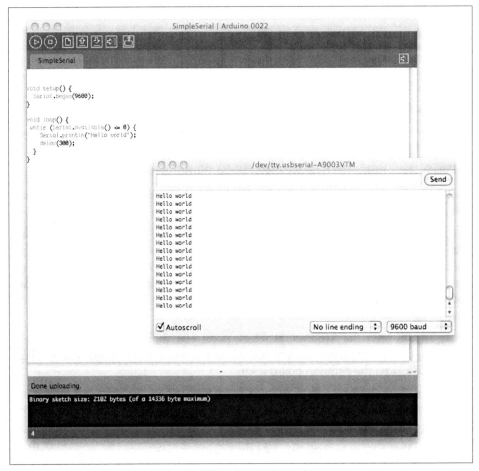

Figure 1-8. The Arduino says "Hello World" in the serial console

Summary

In this chapter we've learned how to use the microcontroller to blink an LED, and then how to get messages from, and send messages to, the board from our Mac. While we will be dealing with sensors later, you can actually work through a fair number of the code examples in this book with just the bare Arduino board, a USB cable, an LED, and the Redpark Serial Cable and RS-232 to TTL serial converter.

CHAPTER 2

Connecting the iPhone to the Arduino

The arrival of the External Accessory framework with iOS 3 was seen, initially at least, as having the potential to open the iOS platform up to a host of external accessories and additional sensors. Sadly, little of the innovation people were expecting actually occurred, and while there are finally starting to be some interesting products arriving on the market, for the most part the External Accessory framework is being used to support a fairly predictable range of audio and video accessories from big-name manufacturers.

The Apple MFi Program

The reason for this lack of innovation is usually laid at the feet of Apple's Made for iPod (MFi) licensing program. The MFi Program is entirely separate from the Apple Developer Program. Being a registered Apple Developer or a member of the iOS Developer Program does not automatically admit you to the MFi Program. However, to develop hardware accessories that connect to the iPod, iPhone, or iPad, you must be an MFi licensee.

 If you are interested in developing your own hardware, more information about the MFi licensing program can be found at *http://developer .apple.com/programs/mfi/*. Licensed developers gain access to technical documentation, hardware components, and technical support that will allow them to build external hardware that can connect to the iPod, iPhone, or iPad.

Unfortunately, becoming a member of the MFi program is not as simple as signing up as an Apple Developer, and it is a fairly lengthy process. From personal experience I can confirm that the process of becoming an MFi licensee is not for the faint-hearted. And once you're a member of the program, getting your hardware out of prototype stage and approved by Apple for distribution and sale is not necessarily a simple process.

The Redpark Serial Cable

Until recently, the restrictions placed on developers by the existence of the MFi program meant that if you were an independent developer, or even a small company, you probably couldn't get access to the documentation and components you needed to connect your iOS device to an arbitrary piece of existing hardware.

Since many of those pieces of existing hardware have serial (RS-232) interfaces, the arrival of the Redpark Serial Cable makes connecting to such hardware fairly easy. The cable arrives in a plain white box; inside is the cable itself (Apple Dock Connector to male 9-pin RS-232 serial) and a loopback test adaptor that allows you to test the cable (see Figure 2-1).

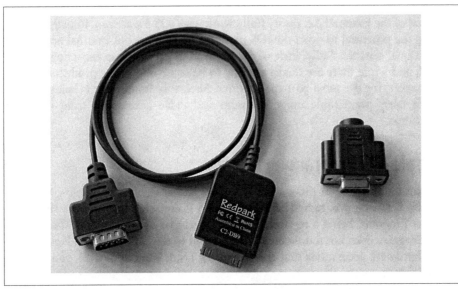

Figure 2-1. The Redpark Serial Cable (left) and loopback adaptor (right)

The Redpark cable is available in the Maker Shed as the "Redpark iPhone Breakout Pack for Arduino" (*https://www.makershed.com/ProductDetails.asp?ProductCode= MSRP02*) or directly from the manufacturer (*http://www.redpark.com/c2db9.html*).

The Redpark Serial Cable SDK comes as a *.dmg* file. You can download the latest version of the SDK from *http://www.redpark.com/c2db9 _Downloads.html*.

Testing the Cable

There are two ways to test that your cable works correctly before we go ahead and start writing code to use it. First, when the cable is plugged in, there should be a new entry in the Settings application; if you go to General→About→Serial Cable, you should see something like Figure 2-2. If the cable is plugged into your iOS device and you do not see this settings entry, or there seems to be something wrong with it, there may be something wrong with your cable.

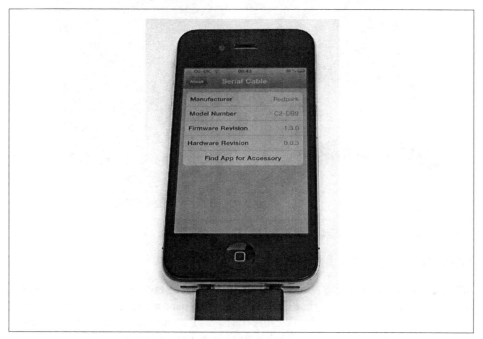

Figure 2-2. Checking the cable in the Settings application

Additionally, the Redpark Serial Cable ships with a loopback adaptor (see Figure 2-3) and a demo application called Rsc Demo. If you open the demo application in Xcode and then build and deploy it onto your device, you should be able to perform a loopback test using this adaptor.

Plug the adaptor into the RS-232 end of the cable, and then plug the cable into the dock connector. Run the Rsc Demo application and tap on the Loop Test button. If you do not see a `UIAlertView`, as in Figure 2-3, informing you that the loopback test has been successful, then, again, there may be something wrong with your cable.

Assuming your new cable passes both of these tests, however, you should assume that all is well, and we can now go ahead and write some code.

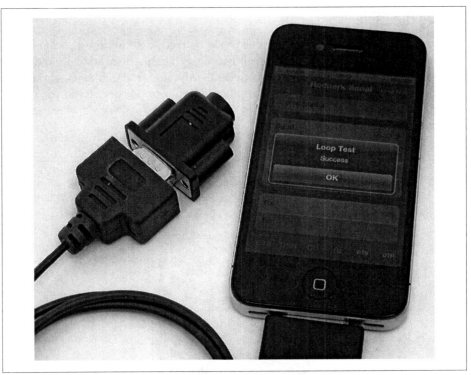

Figure 2-3. Performing a loopback test in the Rsc Demo Application

Connecting to the Arduino

While the original Arduino board used RS-232 serial to connect to the development machine, more recent models use USB. We therefore need an adaptor to translate from RS-232 serial used by the Redpark cable to the TTL serial that we can easily connect to the Arduino transmit/receive (TX/RX) pins (pin 1 and 0, respectively).

To do this conversion I'm using the SparkFun RS-232 Shifter (see Figure 2-4; *http://www.sparkfun.com/products/449*; $13.95), although any equivalent unit would suffice. For instance, the Maker Shed carries the equally effective but slightly cheaper ($7) P4 Adapter Kit (*http://www.makershed.com/ProductDetails.asp?ProductCode=MKMD2*).

As you can see from Figure 2-4, the TTL end has four wires: RX, TX, GND, and VCC (5V). The easiest way to connect this to our Arduino is to solder some snapable header pins to four lengths of single core wire (see Figure 2-5) and build a custom jumper cable (see Figure 2-6). You can, of course, just make use of premade jumper cables; the Maker Shed, for instance, sells a set of Breadboard Jumper Wires (*http://www.makershed.com/ProductDetails.asp?ProductCode=MKSEEED3*) that you could use instead of soldering up your own cable.

Figure 2-4. The SparkFun RS-232 Shifter SMD (sku: PRT-00449)

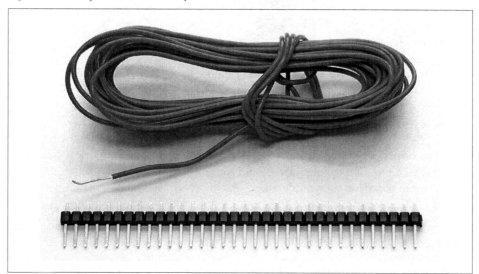

Figure 2-5. Single core wire (top) and snapable header pins (bottom)

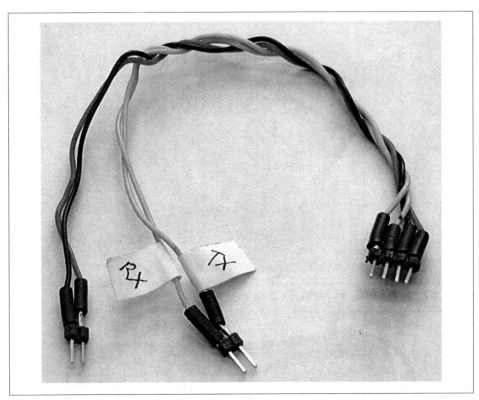

Figure 2-6. The completed jumper cable

Once you have bought or built your jumper cable, connect the Redpark Serial Cable to your RS-232 to TTL serial adaptor, and that in turn (using your jumper cable) to your Arduino board, as shown in Figure 2-7.

To do so, connect the VCC pin of your adaptor to the 5V pin on the Arduino, the GND pin to one of the available GND pins on the Arduino, the RX pin on the adaptor to the TX pin on the Arduino, and the TX pin on the adaptor to the RX pin on the Arduino.

 The receive (RX) pin on your Arduino board (pin 0) must be routed to the transmit (TX) pin on your adaptor, and the transmit (TX) pin on the Arduino board (pin 1) must in turn be connected to the receive (RX) pin on your adaptor. If these are not correctly connected, your applications will not work. Do *not* connect RX to RX and TX to TX.

You can power the Arduino board either via a USB cable, plugged into your Mac or a wall adaptor, or via a battery pack. Since I had one lying around, I used a battery pack, but it's entirely up to you.

Figure 2-7. Connecting the cable to the Arduino

Connecting to an iOS Device

Once you've powered your Arduino, and hence the RS-232 to TTL serial adaptor, the only thing left to do is plug the Redpark Serial Cable into your iOS device (Figure 2-8).

Unfortunately, we're going to spend a lot of time plugging and unplugging our iPhone or iPad from the Redpark cable, as we'll need to plug the device into our Mac to deploy our development code to it.

For now, however, we're good to go. Let's go ahead and write some code to demonstrate the cable.

A Simple Serial Application

Our first application making use of the serial cable will be a simple serial terminal, replicating the functionality of the serial console in the Arduino development environment. Open Xcode, choose to create a new project, select a View-based Application for the iPhone, and when prompted name it "SerialConsole" and save it to the Desktop.

Let's start by building our interface. Click on the *SerialConsoleViewController.xib* file to open it in Interface Builder. Open up the Utilities pane and then drag and drop a UINavigationBar, UITextField, UIButton, and a UITextView into the View from the Object library. Arrange them in a fashion similar to Figure 2-9.

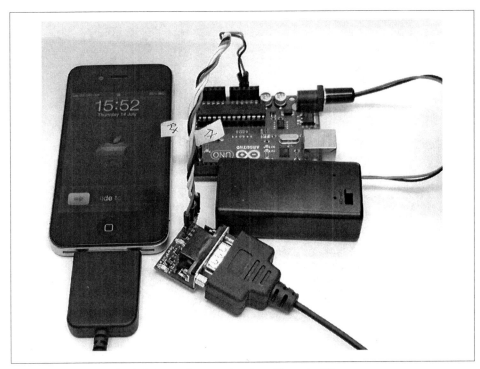

Figure 2-8. Powering the Arduino, and connecting the cable to the iPhone

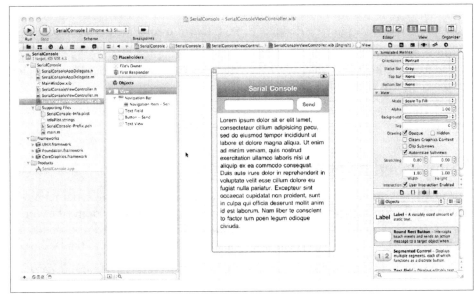

Figure 2-9. Building the SerialConsole app's user interface

We're going to make use of the real estate offered by the navigation bar later in the chapter to add a `UIBarButtonItem`. But for now, change the title text to read "Serial Console," and we'll come back to it later on.

Finally, select and delete the Lorem ipsum text from the `UITextView`, and uncheck the Editable checkbox in the Attributes tab of the Utilities pane so that users won't be able to obscure the view of the text window by accidentally popping up the keyboard.

Close the Utilities pane and open the Assistant Editor, making sure it's set to Automatic and is displaying the *SerialConsoleViewController.h* file.

Then Ctrl-click and drag from the `UITextField`, `UIButton`, and `UITextView` to the editor window to create an `IBOutlet` property for each item, named **textEntry**, **sendButton**, and **serialView**, respectively.

Then Ctrl-click and drag once again from the `UIButton` to create an `IBAction` called `sendString:` (see Figure 2-10). Finally, right-click on the `UITextField` to bring up the Text Fields outlets, events, and actions pane and click and drag to connect the Did End on Exit event to the same `sendString:` method. Doing this means that the same `send String:` method is called either when you tap on the Send button or tap on Return when entering characters into the text entry field itself.

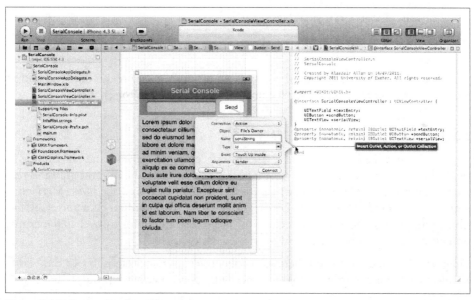

Figure 2-10. Connecting the outlets and actions

After connecting your UI elements to the interface file, your *SerialConsoleViewController.h* header file should look something like this:

```
#import <UIKit/UIKit.h>

@interface SerialConsoleViewController : UIViewController {

    UITextField *textEntry;
    UIButton *sendButton;
    UITextView *serialView;
}

@property (nonatomic, retain) IBOutlet UITextField *textEntry;
@property (nonatomic, retain) IBOutlet UIButton *sendButton;
@property (nonatomic, retain) IBOutlet UITextView *serialView;

- (IBAction)sendString:(id)sender;

@end
```

with corresponding changes in the implementation file. Now click on the *SerialConsoleViewController.m* implementation file, and add the following line to the send String: method:

```
- (IBAction)sendString:(id)sender {
    [self.textEntry resignFirstResponder];

    // Add code here
}
```

This will ensure that the keyboard is dismissed when the user presses the Send button to send a string of characters. We'll come back to this later in the chapter.

Adding the Redpark Serial Library

Now we've got the bare bones of a user interface; let's go ahead and start filling in behind the scenes. Open up your copy of the Redpark Serial SDK and grab the *redparkSerial.h* and *rscMgr.h* header files from the *inc/* folder, and the *libRscMgrUniv.a* static library from *lib/* (see Figure 2-11). The *libRscMgrUniv.a* library is a universal (fat) library supporting the x86 architecture for use in the Simulator as well as the armv6 and armv7 architectures for use on the device hardware itself.

Drag and drop them into your SerialConsole project, remembering to tick the "Copy items into destination group's folder (if needed)" checkbox when prompted.

Since the SDK makes use of the External Accessory framework, we also need to add this to the project at this stage. Click on the project icon at the top of the Project pane in Xcode, then click on the SerialConsole Target, and then on the Build Phases tab. Finally, click on the Link Binary with Libraries item to open up the list of linked frameworks, and click on the + symbol to add a new framework. Select the External Accessory framework from the drop-down list and click the Add button (as in Figure 2-12).

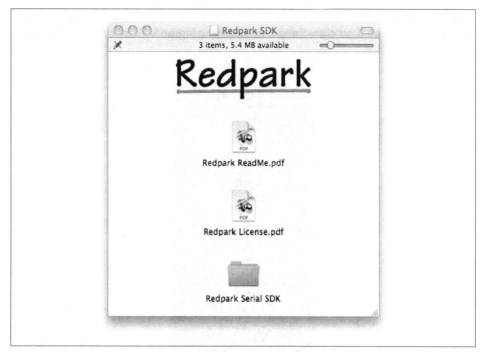

Figure 2-11. The Redpark Serial SDK

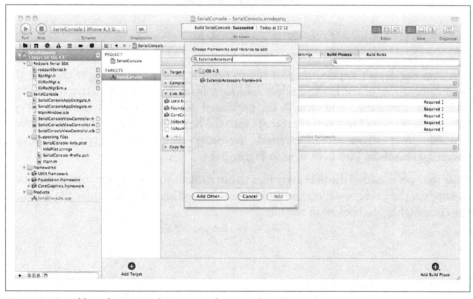

Figure 2-12. Adding the External Accessory framework to the project

Finally, we need to declare support for the cable in our application's *Info.plist* file. If we don't do this, we'll get something that looks like Figure 2-13 every time we plug the cable into the device.

Figure 2-13. The unsupported External Accessory window

Click on the *SerialConsole-Info.plist* file to open it in the editor. Right-click on the bottommost row of the list and select Add Row from the menu. An extra row will be added to the table, and you'll be presented with a drop-down menu. Type **UISupportedExternalAccessoryProtocols** into the box. This will change to the human-readable text "Supported external accessory protocols." Type the string **com.redpark.hobdb9** into Item 0, as in Figure 2-14.

Now that we've added the necessary files to our project and configured our application to support the cable, click on the *SerialConsoleViewController.h* header file and add the code highlighted below to the file:

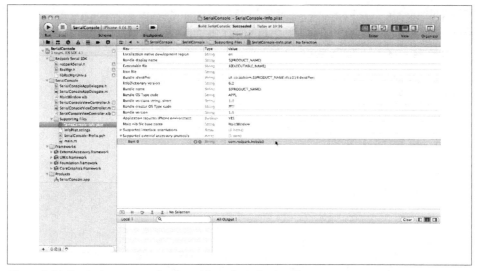

Figure 2-14. Declaring support for the cable in the Info.plist file

```objc
#import <UIKit/UIKit.h>
#import "RscMgr.h"

#define BUFFER_LEN 1024

@interface SerialConsoleViewController : UIViewController <RscMgrDelegate> {

    RscMgr *rscMgr;
    UInt8 rxBuffer[BUFFER_LEN];
    UInt8 txBuffer[BUFFER_LEN];

    UITextField *textEntry;
    UIButton *sendButton;
    UITextView *serialView;
}


@property (nonatomic, retain) IBOutlet UITextField *textEntry;
@property (nonatomic, retain) IBOutlet UIButton *sendButton;
@property (nonatomic, retain) IBOutlet UITextView *serialView;

- (IBAction)sendString:(id)sender;

@end
```

Then, in the corresponding *SerialConsoleViewController.m* implementation file, un-comment the `viewDidLoad` method and add the following highlighted lines:

```objc
- (void)viewDidLoad {
    [super viewDidLoad];
    rscMgr = [[RscMgr alloc] init];
    [rscMgr setDelegate:self];
}
```

This will initialize the Redpark Serial Cable Manager and set its delegate to be this class. After doing so, we need to add the mandatory delegate callbacks:

```
#pragma mark - RscMgrDelegate methods

- (void) cableConnected:(NSString *)protocol {
    [rscMgr setBaud:9600];
    [rscMgr open];
}

- (void) cableDisconnected {

}

- (void) portStatusChanged {

}

- (void) readBytesAvailable:(UInt32)numBytes {
    int bytesRead = [rscMgr read:rxBuffer Length:numBytes];❶
    NSLog( @"Read %d bytes from serial cable.", bytesRead );
    for(int i = 0;i < numBytes;++i) {
        self.serialView.text =
            [NSString stringWithFormat:@"%@%c",
                        self.serialView.text,
                        ((char *)rxBuffer)[i]];❷
    }
}

- (BOOL) rscMessageReceived:(UInt8 *)msg TotalLength:(int)len {
    return FALSE;
}

- (void) didReceivePortConfig {

}
```

❶ Here we pass the rxBuffer instance variable by reference into the rscMgr's read:Length: method, which will modify it in-place. This isn't normal behavior for an Objective-C method, so be careful.

❷ Here we're appending the characters in the newly populated rxBuffer array to the text already in the UITextView.

Although we have to implement all of the mandatory methods, in our bare-bones implementation we're really only interested in the cableConnected: and readBytesAvailable: callbacks.

Connecting the Arduino

We've reached a good point to stop and test the code; click on the Run button in the Xcode toolbar to build and deploy the code onto your iPhone. Once the application

---

has been put onto the iPhone, click on the Stop button to stop it running and unplug your phone from your Mac.

To test the code, we can use the simple serial sketch we used in the previous chapter:

```
void setup() {
  Serial.begin(9600);
}

void loop() {
  while (Serial.available() <= 0) {
    Serial.println("Hello world");
    delay(300);
  }

  Serial.println("Goodbye world");
  while(1) { }
}
```

If you don't already have the sketch loaded onto an Arduino, open up the Arduino development environment and load the sketch. Then connect the Arduino to the Redpark cable, and the Redpark cable to your iPhone. Then tap on the SerialConsole app to resume execution.

If you've plumbed everything together correctly, you should get something that looks very much like Figure 2-15. Just like the serial console on your Mac, the Arduino says "Hello World."

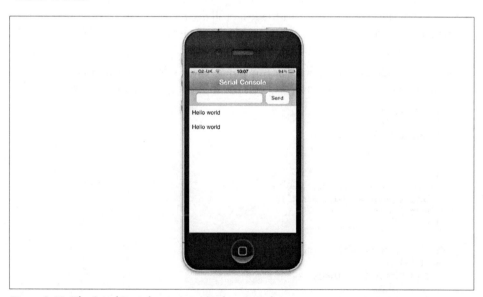

Figure 2-15. The SerialConsole app connected to an Arduino

Sending Data Back to the Arduino

While your iPhone can now receive data from the Arduino, we can't yet send data. Don't worry; this is actually a lot easier than receiving data. Click on the *SerialConsoleViewController.m* implementation file and scroll down until you find the send String: method. Add the following lines:

```
- (IBAction)sendString:(id)sender {
    [self.textEntry resignFirstResponder];

    NSString *text = self.textEntry.text;
    int bytesToWrite = text.length;
    for ( int i = 0; i < bytesToWrite; i++ ) {
        txBuffer[i] = (int)[text characterAtIndex:i];
    }
    int bytesWritten = [rscMgr write:txBuffer Length:bytesToWrite];
}
```

Here we take the text in the UITextField widget and pack the string into the txBuffer one byte at a time. We then use the rscMgr and write the buffer to the serial port.

Go ahead and unplug your iPhone from the Redpark cable, and plug it back into your Mac. Once it's visible again to Xcode, click on the Run button in Xcode to build and deploy the application to your iPhone once again. After it starts up, click on the Stop button and disconnect the iPhone from your Mac and reconnect the Redpark cable. Tap on the SerialConsole application to resume execution.

When you see the "Hello World" messages scroll up the text view, tap on the text entry widget and type a few characters. Then tap the Send button. You should see something like Figure 2-16, and the "Goodbye World" message.

If you want to add a blinking light at this point, you could insert an LED into pin 13, as we did in the previous chapter, and add the following familiar code to our sketch:

```
void setup() {
  Serial.begin(9600);
  pinMode(13, OUTPUT);
}

void loop() {
  while (Serial.available() <= 0) {
    Serial.println("Hello world");
    delay(300);
  }

  Serial.println("Goodbye world");
  while(1) {
    digitalWrite(13, HIGH);
    delay(1000);
    digitalWrite(13, LOW);
    delay(1000);
  }
}
```

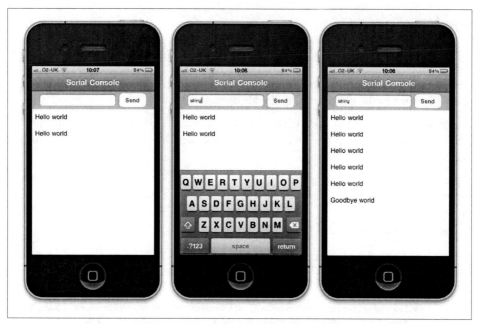

Figure 2-16. Application connected to the Arduino (left), entering characters (middle), which have been received by the Arduino (right)

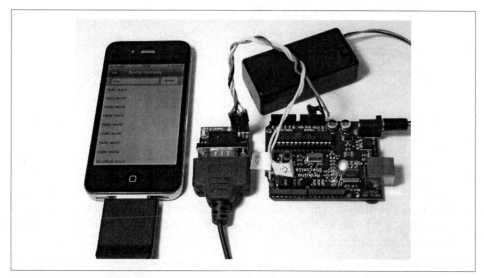

Figure 2-17. The SerialConsole application, cabling, and the Arduino

If you upload this modified sketch to your Arduino board, as well as printing "Goodbye world" when it receives bytes over the serial port, the Arduino will begin to blink the LED on pin 13 on and off once a second. See Figure 2-17.

Log Messages

One of the things that's going to start to irritate you fairly quickly is how hard it is to debug applications using the Redpark cable. With the cable plugged in, you're going to be unable to see output to the debug console: for instance, log messages you might normally print to the console using the NSLog method.

Fortunately, there is a way around this. We can actually redirect the stderr output that would normally appear in the Debug area to a file, which we can then view inside our application itself.

Open up the SerialConsole project in Xcode and click on the *main.m* file, which is located in the Supporting Files group in the Project pane, and add the highlighted lines below to it:

```
#import <UIKit/UIKit.h>

int main(int argc, char *argv[]){
    NSAutoreleasePool *pool = [[NSAutoreleasePool alloc] init];

    NSArray *paths =
      NSSearchPathForDirectoriesInDomains(NSDocumentDirectory, NSUserDomainMask, YES);
    NSString *log = [[paths objectAtIndex:0] stringByAppendingPathComponent:
@"ns.log"];

    NSFileManager *fileMgr = [NSFileManager defaultManager];
    [fileMgr removeItemAtPath:log error:nil];

    freopen([log fileSystemRepresentation], "a", stderr);

    int retVal = UIApplicationMain(argc, argv, nil, nil);
    [pool release];
    return retVal;
}
```

This will redirect the stderr output to a file called *ns.log* inside the application's Documents directory. The file will be deleted and recreated each time the application is started.

Now that we've done this, we need to create a view controller to allow us to look at the log. Right-click on the Project pane and select New File..., select a UIViewController subclass from the drop-down list, which is a subclass of UIViewController itself, and make sure the "With XIB for user interface" checkbox is checked. Name the new class *LogController* when prompted.

Click on the *LogController.xib* file to open it in Interface Build, and drag a UINaviga tionBar and a UIBarButtonItem from the Object library into your view. Position the bar button item on the left of the navigation bar and change the text to be "Done." Then in the Attributes inspector, change the button style to Done rather than Plain. You should also change the title of the navigation bar to "NSLog."

Now drag a `UITextView` from the Object library into the view to fill the remaining space. Delete the Lorem ipsum text, and in the Attributes inspector, uncheck the Editable checkbox.

Next, close the Utilities pane and open the Assistant Editor. Make sure it's on Automatic, and Ctrl-click and drag from the `UITextView` to the displayed *LogController.h* interface file to create a property and `IBOutlet` called `logWindow`. Then, similarly, Ctrl-click and drag from the Done button the interface file to create an `IBAction` called `done:`; see Figure 2-18.

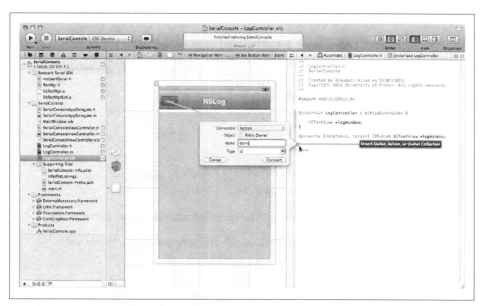

Figure 2-18. Connecting the outlets and actions to the user interface

Save your changes, return to the Standard Editor, and click on the *LogController.h* interface file to open it in the Xcode editor. Add the highlighted lines below:

```
#import <UIKit/UIKit.h>

@interface LogController : UIViewController {

    UITextView *logWindow;
    BOOL firstOpen;

}

@property (nonatomic, retain) IBOutlet UITextView *logWindow;

- (IBAction)done:(id)sender;
- (void)setWindowScrollToVisible;

@end
```

Then, click on the corresponding *LogController.m* implementation file. The first thing we need to do is modify the `initWithNibName:` method as follows:

```
- (id)initWithNibName:(NSString *)nibNameOrNil bundle:(NSBundle *)nibBundleOrNil {
    self = [super initWithNibName:nibNameOrNil bundle:nibBundleOrNil];
    if (self) {
        NSArray *paths = NSSearchPathForDirectoriesInDomains(
            NSDocumentDirectory, NSUserDomainMask, YES);
        NSString *log =
            [[paths objectAtIndex:0] stringByAppendingPathComponent: @"ns.log"];
        NSFileHandle *fh = [NSFileHandle fileHandleForReadingAtPath:log];

        [fh seekToEndOfFile];

        [[NSNotificationCenter defaultCenter] addObserver:self
                                    selector:@selector(getData:)
                                    name:@"NSFileHandleReadCompletionNotification"
                                    object:fh];
        [fh readInBackgroundAndNotify];
        firstOpen = YES;
    }
    return self;
}
```

From there we need to add the `viewDidAppear:` method:

```
- (void)viewDidAppear:(BOOL)animated {
    [super viewDidAppear:animated];
    NSArray *paths = NSSearchPathForDirectoriesInDomains(NSDocumentDirectory,
NSUserDomainMask, YES);
    NSString *log = [[paths objectAtIndex:0] stringByAppendingPathComponent:
@"ns.log"];

    if ( firstOpen ) {
        NSString* content = [NSString stringWithContentsOfFile:log
encoding:NSUTF8StringEncoding error:NULL];
        logWindow.editable = TRUE;
        logWindow.text = [logWindow.text stringByAppendingString: content];
        logWindow.editable = FALSE;
        firstOpen = NO;
    }
}
```

Then we need to add the `getData:` method called when the class is notified there is a change in the log file, along with some supporting methods:

```
#pragma mark - NSLog Redirection Methods

- (void) getData: (NSNotification *)aNotification {
    NSData *data = [[aNotification userInfo]
objectForKey:NSFileHandleNotificationDataItem];
    if ([data length]) {
        NSString *aString = [[[NSString alloc] initWithData:data
encoding:NSUTF8StringEncoding] autorelease];

        logWindow.editable = TRUE;
```

```
        logWindow.text = [logWindow.text stringByAppendingString: aString];
        logWindow.editable = FALSE;

        [self setWindowScrollToVisible];
        [[aNotification object] readInBackgroundAndNotify];
    } else {
        [self performSelector:@selector(refreshLog:) withObject:aNotification
afterDelay:1.0];
    }
}
- (void) refreshLog: (NSNotification *)aNotification {
    [[aNotification object] readInBackgroundAndNotify];
}

-(void)setWindowScrollToVisible {
    NSRange txtOutputRange;
    txtOutputRange.location = [[logWindow text] length];
    txtOutputRange.length = 0;
    logWindow.editable = TRUE;
    [logWindow scrollRangeToVisible:txtOutputRange];
    [logWindow setSelectedRange:txtOutputRange];
    logWindow.editable = FALSE;
}
```

Finally, you should modify the done: method so that it dismisses the view controller:

```
-(IBAction)done:(id)sender{
    [self dismissModalViewControllerAnimated:YES];
}
```

Almost done. Save your changes and click on the *SerialConsoleViewController.xib* file
to open it in Interface Builder. Drag and drop a UIBarButtonItem from the Object library
into the navigation bar, on the left of the title, and change the name to Log.

Close the Utilities pane and switch to the Assistant Editor, making sure it's in Automatic
mode, and Ctrl-click and drag from the Log button to the *SerialConsoleViewControl-
ler.h* interface file to create an openLog: action, as in Figure 2-19.

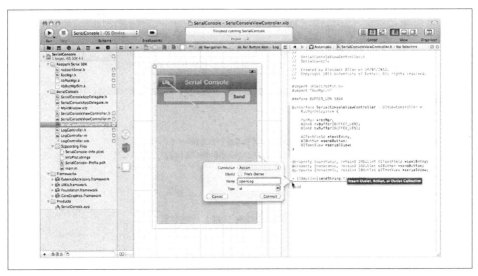

Figure 2-19. Connecting the Log button to the openLog: action

Save your changes, switch to the Standard editor and click on the *SerialConsoleView-Controller.h* interface file, and add the following instance variable:

```
#import <UIKit/UIKit.h>
#import "LogController.h"
#import "RscMgr.h"

#define BUFFER_LEN 1024

@interface SerialConsoleViewController : UIViewController <RscMgrDelegate> {

    RscMgr *rscMgr;
    UInt8 rxBuffer[BUFFER_LEN];
    UInt8 txBuffer[BUFFER_LEN];

    UITextField *textEntry;
    UIButton *sendButton;
    UITextView *serialView;
    UIBarButtonItem *openLog;

    LogController *logWindow;
}

@property (nonatomic, retain) IBOutlet UITextField *textEntry;
@property (nonatomic, retain) IBOutlet UIButton *sendButton;
@property (nonatomic, retain) IBOutlet UITextView *serialView;
@property (nonatomic, retain) LogController *logWindow;

- (IBAction)sendString:(id)sender;
- (IBAction)openLog:(id)sender;

@end
```

Then, click on the *SerialConsoleViewController.m* implementation file to open it in the Xcode editor, and remember to synthesize our new property:

```
@synthesize logWindow;
```

Modify the `viewDidLoad:` method as below:

```
- (void)viewDidLoad {
    [super viewDidLoad];
    rscMgr = [[RscMgr alloc] init];
    [rscMgr setDelegate:self];

    self.logWindow = [[LogController alloc] initWithNibName:@"LogController"
bundle:nil];
}
```

and the `openLog:` method to display the log view controller when the Log button is tapped:

```
- (IBAction)openLog:(id)sender {
    [self presentModalViewController:self.logWindow animated:YES];
}
```

Save all your changes, and click on the Run button in the Xcode toolbar to build and deploy the code onto your device. Before you do that, make sure that your iPhone is plugged into your Mac, and not the Redpark cable.

Once the application is loaded onto your iPhone, click on the Stop button in the Xcode toolbar and disconnect the phone. Restart the application and click on the Log button in navigation bar. If everything works as it should, you should see something along the lines of the following:

```
2011-07-15 15:56:06.542 SerialConsole[4686:707] ACC Connect
```

which is the Redpark SDK telling you that it has initialized itself. At this point, you might want to go through the code and instrument up your delegate methods with a few `NSLog` statements and redeploy the code to your device. If you do that, you might see something like Figure 2-20.

Summary

In this chapter we've built a simple serial console for iOS, and passed messages back and forth between our iPhone and our Arduino board. We've also built a log window view controller that we can reuse in other projects, which will be useful when we're debugging.

In the next chapter, we're going to build something a bit more ambitious, this time for the iPad.

Figure 2-20. The log viewer in action

CHAPTER 3

Controlling the Arduino from the iPad

In the previous chapter, we passed messages back and forth between the Arduino and the iPhone using the Redpark Serial Cable and the SDK. Now let's make the Arduino do something other than sending messages, and turn some LEDs on and off directly from our iPad.

An Arduino on Your iPad

We're going to go ahead and build a simple application for the iPad that will let you pull an arbitrary pin on the Arduino HIGH or set it LOW. While we'll be building the application for the iPad, because I want to take advantage of all that screen real estate, the code will run just fine on the iPhone (or the iPod touch) if you want to build it as an application for the iPhone without any modification.

Open Xcode, choose to create a new project, select a View-based application for the iPad device family, and when prompted name it "Paduino" and save it to the Desktop. When the project opens, deselect Landscape Left and Landscape Right as supported orientations. We're going to be working in portrait mode.

Adding the Serial Library

This time, let's start by adding the Redpark library. Open up your copy of the Redpark Serial SDK, grab the *redparkSerial.h* and *rscMgr.h* header files from the *inc/* folder and the *libRscMgrUniv.a* static libraries from *lib/*, and then drag and drop them into your new Paduino project, remembering to tick the "Copy items into destination group's folder (if needed)" checkbox when prompted.

We also need to add the External Accessories framework, so click on the project icon at the top of the Project pane in Xcode and click on the Paduino Target, and then on the Build Phases tab. Finally, click on the Link Binary with Libraries item to open up the list of linked frameworks, and click on the + symbol to add a new framework. Select the External Accessory framework from the drop-down list, and click the Add button.

Finally, we need to declare support for the cable. Click on the *SerialConsole-Info.plist* file to open it in the editor. Right-click on the bottommost row of the list and select Add Row from the menu. An extra row will be added to the table, and you'll be presented with a drop-down menu. Type **UISupportedExternalAccessoryProtocols** into the box. This will change to the human-readable text "Supported external accessory protocols." Type the string **com.redpark.hobdb9** into Item 0 to declare our support for the cable.

Building the User Interface

The first thing we're going to need is a high-resolution picture of an Arduino. There are plenty of these available on the Web, or you can use an image of your own.

 A good image of the Arduino Uno can be found on the Arduino Project website (*http://arduino.cc/en/uploads/Main/ArduinoUnoFront.jpg*).

If you need to, rotate the image using the Preview application so that the power and USB connections are at the top of the frame, and save your changes. Drag and drop your image into the project.

Now that we've done that, let's build our user interface before going ahead and wiring it up to our view controller. Click on the *PaduinoViewController.xib* file to open it in Interface Builder. Drag and drop a **UINavigationBar** from the Object library and position it at the top of the view. Change the title to be "Paduino." Then drag a **UIImageView** from the library and position it so it fills the remaining view.

In the Attributes inspector, set the View mode to be Aspect Fill, and the Image to be the Arduino image you added to your project a few moments ago. You should see something like Figure 3-1.

Next drag twelve **UISwitch** elements from the Object library and position them as in Figure 3-2, in two blocks of six switches next to digital pins 2 through 13. We're going to use these switches to switch the Arduino's digital pins from LOW to HIGH and back again.

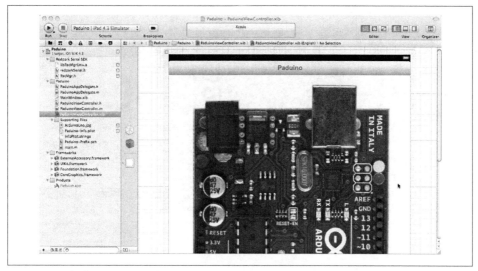

Figure 3-1. The basic user interface

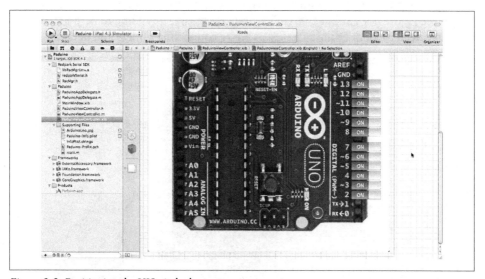

Figure 3-2. Positioning the UISwitch elements

For each of the switches, in the Attributes inspector, set the switch state to be off. This will correspond to holding the Arduino pin LOW; we're going to write our Arduino code so this will be the initial state. Then, in the same inspector window, set the view's tag to be the same as the pin number, see Figure 3-3.

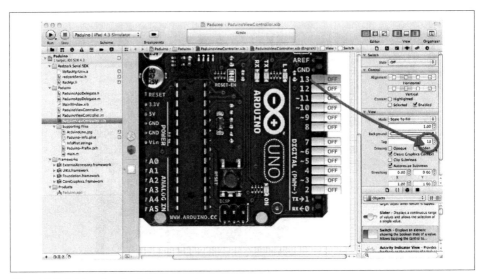

Figure 3-3. Setting the view's tag property

Having done that, flip to the Identity inspector and set the switch label to be the pin name (see Figure 3-4). While this isn't strictly necessary, it is going to help us out later when we need to identify the switches as we're connecting outlets and actions.

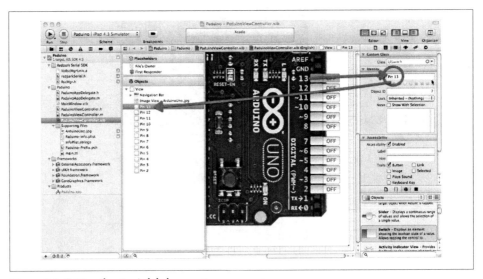

Figure 3-4. Setting the view's label property

Now that we've done that, close the Utilities panel and open the Assistant Editor. Make sure it's on Automatic and Ctrl-click and drag to the *PaduinoViewController.h* interface file from each of the switches to create an instance variable and property for each.

After doing that, click on one of the switches to select it, then Ctrl-click and drag to the interface file to create an `IBAction` called `toggle:` (see Figure 3-5). After doing this, Ctrl-click and drag from all of the other switches to the same method. Changing the state of any of the switches will call this single method, passing the `UISwitch` in as the sender.

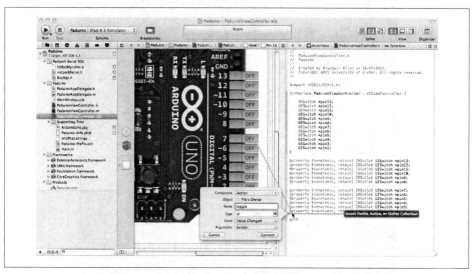

Figure 3-5. Connecting all twelve switches to the toggle: method

After doing this, if you right-click on File's Owner in the object's view, you should see something like Figure 3-6. This pop up shows the connections from File's Owner, which in this case is the *PaduinoViewController*, to the various elements in the view.

We're done for now. We've made a lot of changes, so make sure you've saved, and click on the *PaduinoViewController.h* interface file and open it in the Standard Editor.

Integrating the Serial Library

If you were following along last chapter, closely integrating out user interface and the serial library should look very familiar. In fact, you might even be able to save yourself some time using copy and paste. If you do, remember to be careful; copy and paste errors can get irritating.

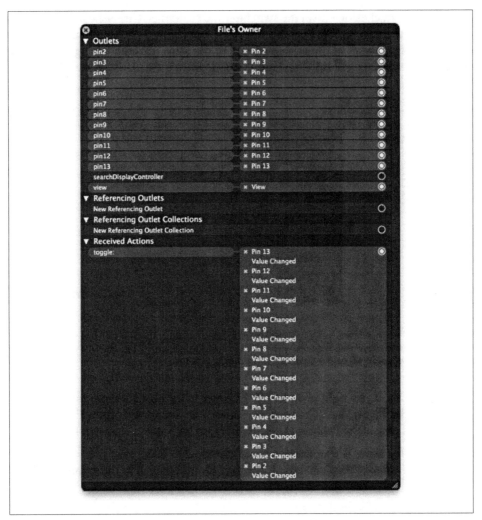

Figure 3-6. Connections from File's Owner to the user interface elements

Add the following highlighted lines to the *PaduinoViewController.h* interface file:

```
#import <UIKit/UIKit.h>
#import "RscMgr.h"

#define BUFFER_LEN 1024

@interface PaduinoViewController : UIViewController <RscMgrDelegate> {

    UISwitch *pin13;
    UISwitch *pin12;
    UISwitch *pin11;
    UISwitch *pin10;
    UISwitch *pin9;
    UISwitch *pin8;
    UISwitch *pin7;
    UISwitch *pin6;
    UISwitch *pin5;
    UISwitch *pin4;
    UISwitch *pin3;
    UISwitch *pin2;

    RscMgr *rscMgr;
    UInt8 rxBuffer[BUFFER_LEN];
    UInt8 txBuffer[BUFFER_LEN];

}

@property (nonatomic, retain) IBOutlet UISwitch *pin13;
@property (nonatomic, retain) IBOutlet UISwitch *pin12;
@property (nonatomic, retain) IBOutlet UISwitch *pin11;
@property (nonatomic, retain) IBOutlet UISwitch *pin10;
@property (nonatomic, retain) IBOutlet UISwitch *pin9;
@property (nonatomic, retain) IBOutlet UISwitch *pin8;

@property (nonatomic, retain) IBOutlet UISwitch *pin7;
@property (nonatomic, retain) IBOutlet UISwitch *pin6;
@property (nonatomic, retain) IBOutlet UISwitch *pin5;
@property (nonatomic, retain) IBOutlet UISwitch *pin4;
@property (nonatomic, retain) IBOutlet UISwitch *pin3;
@property (nonatomic, retain) IBOutlet UISwitch *pin2;

- (IBAction)toggle:(id)sender;

@end
```

Save your changes and switch to the corresponding implementation file, uncomment the viewDidLoad: method, and initialize the Redpark Serial Cable Manager as below:

```
- (void)viewDidLoad {
    [super viewDidLoad];
    rscMgr = [[RscMgr alloc] init];
    [rscMgr setDelegate:self];
}
```

Having done that, add the following mandatory delegates to your implementation:

```
- (void) cableConnected:(NSString *)protocol {
    [rscMgr setBaud:9600];
    [rscMgr open];

}

- (void) cableDisconnected {

}

- (void) portStatusChanged {

}

- (void) readBytesAvailable:(UInt32)numBytes {
    int bytesRead = [rscMgr read:rxBuffer Length:numBytes];
    NSLog( @"Read %d bytes from serial cable.", bytesRead );

    NSString *string = nil;
    for(int i = 0;i < numBytes;++i) {
        if ( string ) {
            string = [NSString stringWithFormat:@"%@%c", string, rxBuffer[i]];
        } else {
            string = [NSString stringWithFormat:@"%c", rxBuffer[i]];
        }
    }
    NSLog(@"Received: %@", string);
}

- (BOOL) rscMessageReceived:(UInt8 *)msg TotalLength:(int)len {
    return FALSE;
}

- (void) didReceivePortConfig {

}
```

All of this should look fairly familiar from the previous chapter. So let's go ahead and look at our **toggle:** method:

```
- (IBAction)toggle:(id)sender {

    NSLog(@"Toggled output pin %i to %i", [sender tag], [(UISwitch *)sender isOn] );
    txBuffer[0] = [sender tag];
    txBuffer[1] = [(UISwitch *)sender isOn];
    int bytesWritten = [rscMgr write:txBuffer Length:2];
    NSLog( @"Wrote %d bytes to serial cable.", bytesWritten);
}
```

Despite being short, this is the core of our application. Do you remember that while we were building out our interface I ensured that each **UISwitch** had a unique tag corresponding to the switch number? Here we're using that tag to build a two-byte message to the Arduino board. The first byte is the switch number and the second byte is a 1 or a 0, corresponding to the state of the switch, on or off. On the Arduino side of things, that's going to correspond to a pin state of HIGH or LOW.

We're done, at least for now, with the code for the iPad. Make sure all your changes are saved, and click on the Run button in the Xcode toolbar to build and deploy your application onto your iPad. Once it has been successfully loaded, click on the Stop button and put your iPad to one side.

Listening for Messages on the Arduino

Open up the Arduino development environment, enter the following code, and upload it to your Arduino board (if you're unsure how to do that, have a look at Chapters 1 and 2 again):

```
#define LENGTH 2

int rxBuffer[128];
int rxIndex  = 0;

void setup() {

  Serial.begin(9600);
  Serial.println("Arduino reset");

  pinMode(2, OUTPUT);
  pinMode(3, OUTPUT);
  pinMode(4, OUTPUT);
  pinMode(5, OUTPUT);
  pinMode(6, OUTPUT);
  pinMode(7, OUTPUT);
  pinMode(8, OUTPUT);
  pinMode(9, OUTPUT);
  pinMode(10, OUTPUT);
  pinMode(11, OUTPUT);
  pinMode(12, OUTPUT);
  pinMode(13, OUTPUT);

  digitalWrite( 12, HIGH);
  digitalWrite( 13, HIGH);
  delay(500);
  digitalWrite( 12, LOW );
  digitalWrite( 13, LOW );
}

void loop (){

 if (Serial.available() > 0) {

   rxBuffer[rxIndex++] = Serial.read();❶
   if (rxIndex == LENGTH) {❷

       Serial.print( "Set: " );
       Serial.print( rxBuffer[0], DEC );
```

```
            Serial.print( " to: " );
            Serial.println( rxBuffer[1], DEC );

            if( rxBuffer[1] == 0 ) {
              digitalWrite((int)rxBuffer[0], LOW);❸
            } else {
              digitalWrite((int)rxBuffer[0], HIGH);❹
            }
            rxIndex = 0;
        }
    }
    delay(10);

    }
```

❶ We have bytes available, so read a single byte from the serial buffer and put it into the rxBuffer array at the next available free index. Then post-increment our index into the buffer.

❷ If we have received LENGTH bytes (in this case LENGTH is 2 bytes), we have the entire message from the remote host and we should deal with it now. If not, loop and read more bytes from the serial buffer.

❸ If the second byte was a 0, we're going to set the pin to LOW. Read the first byte of the buffer, as that's the pin number, and pull that pin LOW.

❹ Alternatively, if the second received byte was a 1, we're going to set the pin to HIGH. Read the first byte of the buffer, as that's the pin number, and pull that pin HIGH.

Unless we have a voltmeter handy, there isn't any way to know if a pin on the Arduino board is LOW or HIGH, so we're going to need some way to tell. An easy way to do this is to put an LED on each pin. We could wire each pin separately, but an easy hack is shown in Figure 3-7. Here I've wired the ground legs of three LEDs together to a jumper wire that we can then insert in the GND pin of the Arduino board.

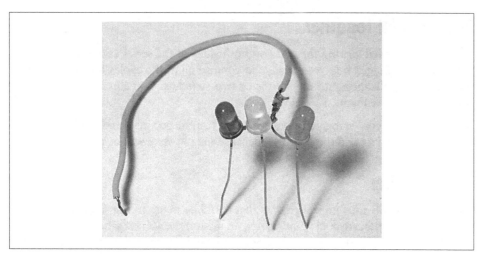

Figure 3-7. Wiring some LEDs together

The three other legs will go into pins 13, 12, and 11, respectively; see Figure 3-8.

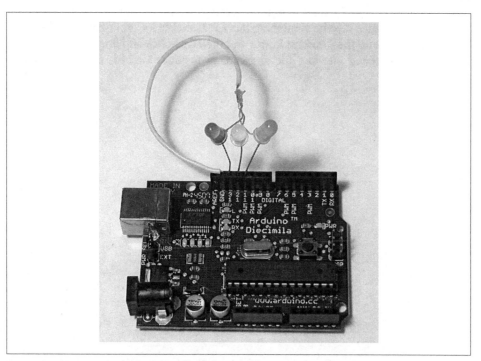

Figure 3-8. Connecting the LED bundle to the Arduino

Putting It All Together

We've now finished writing the code on both sides of the serial connection. Unplug whatever was plugged into your Mac, and connect all of the cables together. Start up the Paduino application and toggle one or two switches. You should see something that looks a lot like Figure 3-9 below.

Remember, once you've toggled a pin to HIGH, you can go ahead and unplug the serial cable from your iPad and the Arduino, because the pin state will be persistent.

Going Further

There are some obvious improvements that could be made to our application. Right now, there is no indication of what's going on behind the scenes, which makes debugging difficult. Let's start to fix that by adding a label to our user interface and hooking this up to the `cableConnected` and `cableDisconnected` delegate methods so we know for sure whether the cable is properly connected to our iPad.

Click on the *PaduinoViewController.xib* interface file to open it in Interface Builder and drag and drop two `UILabel` elements into the view. Position them at the top of your interface and in the top left corner. Change the text of the first label to be "Cable:" and position the second to the right of that. You can leave the second label with its default text intact, or change it to be something more reasonable for an initial state (like Disconnected). Open up the Assistant Editor and then Ctrl-click and drag from the second label to the associated interface file to create an instance variable and property (see Figure 3-10).

Figure 3-9. The iPad connected to the Arduino using the Redpark cable (top), and a close up of the UISwitch positions and LEDs on the Arduino board (bottom)

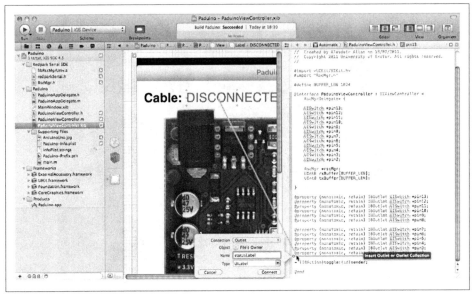

Figure 3-10. Connecting the UILabel to the interface file

Save your changes, then switch to the Standard Editor and return to the *PaduinoView-Controller.m* implementation file. Go to the `cableConnected` and `cableDisconnected` delegate methods and add the highlighted code:

```
- (void) cableConnected:(NSString *)protocol {
    [rscMgr setBaud:9600];
    [rscMgr open];
    self.statusLabel.text = @"CONNECTED";
    self.statusLabel.textColor = [UIColor greenColor];
}

- (void) cableDisconnected {
    self.statusLabel.text = @"DISCONNECTED";
    self.statusLabel.textColor = [UIColor redColor];
}
```

This should change the status text and color when we connect and disconnect the cable.

Adding a Log Window

It's actually fairly easy to reuse the log window we wrote in the previous chapter.

Open the SerialConsole from the previous chapter, and select the *LogController.h*, *LogController.m*, and *LogController.xib* files from the Project pane. Drag and drop them into the Paduino project, remembering to check the "Copy files into destination group's folder (if needed)" checkbox.

Once you've done that, open the *PaduinoViewController.xib* file in Interface Builder and drag and drop a `UIBarButtonItem` from the Object library into the navigation bar. Change the title of the button to "Log." Close the Utilities pane and open up the Assistant Editor. Then Ctrl-click and drag from the Log button into the associated header file in the editor to create an `IBAction` method called `openLog:` (see Figure 3-11).

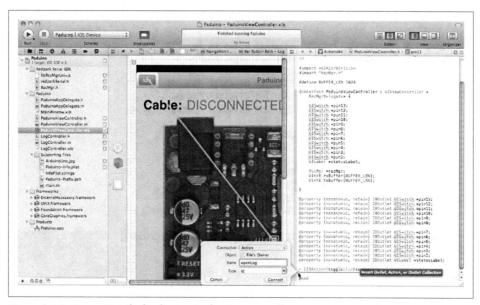

Figure 3-11. Connecting the log button to the view controller.

Switch back to the standard editor, save your changes, and then click on the *Paduino-ViewController.h* file to open it in the editor. Add the highlighted code below:

```
#import <UIKit/UIKit.h>
#import "RscMgr.h"
#import "LogController.h"

#define BUFFER_LEN 1024

@interface PaduinoViewController : UIViewController <RscMgrDelegate> {

    UISwitch *pin13;
    UISwitch *pin12;
    UISwitch *pin11;
    UISwitch *pin10;
    UISwitch *pin9;
    UISwitch *pin8;
    UISwitch *pin7;
    UISwitch *pin6;
    UISwitch *pin5;
    UISwitch *pin4;
    UISwitch *pin3;
```

```
        UISwitch *pin2;
        UILabel *statusLabel;

        RscMgr *rscMgr;
        UInt8 rxBuffer[BUFFER_LEN];
        UInt8 txBuffer[BUFFER_LEN];

        LogController *logWindow;
    }

    @property (nonatomic, retain) IBOutlet UISwitch *pin13;
    @property (nonatomic, retain) IBOutlet UISwitch *pin12;
    @property (nonatomic, retain) IBOutlet UISwitch *pin11;
    @property (nonatomic, retain) IBOutlet UISwitch *pin10;
    @property (nonatomic, retain) IBOutlet UISwitch *pin9;
    @property (nonatomic, retain) IBOutlet UISwitch *pin8;

    @property (nonatomic, retain) IBOutlet UISwitch *pin7;
    @property (nonatomic, retain) IBOutlet UISwitch *pin6;
    @property (nonatomic, retain) IBOutlet UISwitch *pin5;
    @property (nonatomic, retain) IBOutlet UISwitch *pin4;
    @property (nonatomic, retain) IBOutlet UISwitch *pin3;
    @property (nonatomic, retain) IBOutlet UISwitch *pin2;
    @property (nonatomic, retain) IBOutlet UILabel *statusLabel;

    @property (nonatomic, retain) IBOutlet LogController *logWindow;

    - (IBAction)toggle:(id)sender;
    - (IBAction)openLog:(id)sender;

    @end
```

Then, switching to the corresponding *PaduinoViewController.m* implementation file, you should remember to:

```
    @synthesize logWindow;
```

before going on to add the following code to initialize the log window in the `viewDid Load:` method:

```
        self.logWindow = [[LogController alloc] initWithNibName:@"LogController"
                                                         bundle:nil];
```

and release it in the `dealloc` method:

```
        [logWindow release];
```

Finally, in the `openLog:` method you should modally present the log controller:

```
    - (IBAction)openLog:(id)sender {
        [self presentModalViewController:self.logWindow animated:YES];
    }
```

Save your changes and then click on the *main.m* file, which is located in the Supporting Files group in the Project pane. As we did in the previous chapter, add the highlighted lines below to it:

```
#import <UIKit/UIKit.h>

int main(int argc, char *argv[]){
    NSAutoreleasePool *pool = [[NSAutoreleasePool alloc] init];

    NSArray *paths =
      NSSearchPathForDirectoriesInDomains(NSDocumentDirectory, NSUserDomainMask, YES);
    NSString *log = [[paths objectAtIndex:0] stringByAppendingPathComponent:
                       @"ns.log"];

    NSFileManager *fileMgr = [NSFileManager defaultManager];
    [fileMgr removeItemAtPath:log error:nil];

    freopen([log fileSystemRepresentation], "a", stderr);

    int retVal = UIApplicationMain(argc, argv, nil, nil);
    [pool release];
    return retVal;
}
```

This will redirect the stderr output to a file called *ns.log* inside the application's Documents directory.

At this stage, you should go ahead and add NSLog messages in various vital places, such as the RscMgr delegate methods, to report what's going on inside your application. If you now save your changes, and click on the Run button in the Xcode editor to build and deploy the application to your device, you should have a lot more feedback when you plug and unplug the cable.

Summary

In this chapter, we built a simple application to control your Arduino directly from your iPad. In the next chapter we'll go a step further and start gathering sensor measurements.

CHAPTER 4

Using External Sensors from the iPhone

So far we've successfully passed messages back and forth between the Arduino board and our iPhone, and then gone on to control the board directly from our iPad. In this chapter, we're going to take another step and build an application that can turn a remote sensor on, parse the values it gets from the sensor, and then plot the resulting measurements, all in real time.

The LV-MaxSonar-EZ1

The sensor we're going to be using in this chapter is the LV-MaxSonar-EZ1 manufactured by MaxBotics (see Figure 4-1), operating at 5 or 3.3 V.

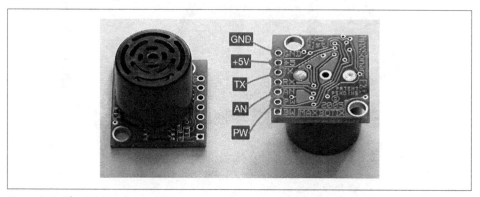

Figure 4-1. The LV-MaxSonar-EZ1

The EZ1 is a versatile ultrasonic range finder and is widely available from various suppliers, including SparkFun (*http://www.sparkfun.com/products/639*), for around $25.

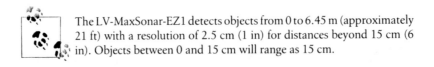 The LV-MaxSonar-EZ1 detects objects from 0 to 6.45 m (approximately 21 ft) with a resolution of 2.5 cm (1 in) for distances beyond 15 cm (6 in). Objects between 0 and 15 cm will range as 15 cm.

Interestingly for our purposes, the sensor provides three different interfaces: analog voltage output, pulse width output, and serial digital output. All three interfaces are active simultaneously. For more information see the product datasheet (*http://www.maxbotix.com/documents/MB1010_Datasheet.pdf*).

Analog Output

Perhaps the easiest, and most familiar, way to wire the EZ1 sensor is to make use of its analog output. We need only three wires to the sensor, connected to the GND, +5V, and AN pins. See Figure 4-1 and Figure 4-2 for details.

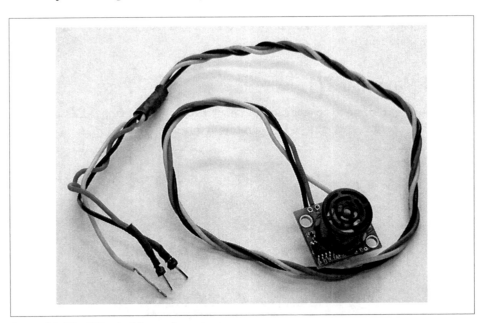

Figure 4-2. The EZ1 wired for analog output

Making use of the EZ1 sensor operating in this mode is fairly trivial. If you connect the 5V and GND pins to their respective counterparts on the Arduino board, and the AN pin to Analog pin A0, the following sketch will send distance measurements to the Arduino's serial pins (RX, TX) and to the USB connection:

```
int ez1Analog = 0; // Analog Pin, A0

void setup() {
```

```
      pinMode(ez1Analog,INPUT);
      Serial.begin(9600);
   }

void loop() {
      int val = analogRead(ez1Analog);
      if (val > 0) {
         val = val / 2;
         float cm = float(val)*2.54;
         Serial.println( int(cm) ); // cm
      }
   }
```

Pulse Width Output

An alternative to using the AN pin on the EZ1 sensor is to use the PW pin and make use of the pulse width output.

The pulse width is 58 ms per cm, or 147 ms per in.

If you disconnect the AN pin, and connect the EZ1's PW pin to digital pin 5 on the Arduino, which is one of the pins that supports PWM, the following sketch replicates the functionality of the sketch in the preceding section in a slightly different fashion:

```
int ez1Pulse = 5; // Digital Pin 5 (with PWM)

void setup() {
   pinMode(ez1Pulse,INPUT);
   Serial.begin(9600);
}

void loop() {
   int val = pulseIn(ez1Pulse, HIGH);
   if (val > 0) {
      float cm = val / 58; // pulse width is 58 microsecs per cm
      Serial.println( int(cm) ); // cm
   }
}
```

RS-232 Serial Output

One of the main reasons I selected the EZ1 as our example sensor for this chapter is that this sensor offers asynchronous serial reporting of measurements in addition to analog and pulse width outputs. This means that we can bypass the Arduino board and the RS-232 to TTL adaptor, and connect the sensor directly to the Redpark cable. We'll look at how to do that toward the end of the chapter.

The TX output delivers asynchronous serial with an RS-232 format (although voltage range is 0–Vcc). Output is an ASCII capital "R", followed by three ASCII character digits representing the range in inches (up to a maximum of 255), followed by a carriage return (ASCII character 13).

Although the voltage of 0–Vcc is outside the RS-232 standard, most RS-232 devices have sufficient margin to read 0–Vcc serial data. The baud rate is 9600, 8 bits, no parity, with one stop bit.

MaxSonar Range Finder for iPhone

At least in the first instance, we're going to make use of the sensor in analog mode, and connect it to our Arduino board, which in turn will be connected to our iPhone via the Redpark cable.

Open Xcode, choose to create a new project, select a View-based application for the iPhone device family, and when prompted name it "MaxSonar" and save it to the Desktop.

Adding the Serial Library

Open your copy of the Redpark Serial SDK and grab the *redparkSerial.h* and *rscMgr.h* header files from the *inc/* folder, and the *libRscMgrUniv.a* static libraries from *lib/*. Drag and drop them into your new MaxSonar project, remembering to tick the "Copy items into destination group's folder (if needed)" checkbox when prompted.

We also need to add the External Accessories framework, so click on the project icon at the top of the Project pane in Xcode, and click on the MaxSonar Target and then on the Build Phases tab. Finally, click on the Link Binary with Libraries item to open up the list of linked frameworks, and click on the + symbol to add a new framework. Select the External Accessory framework from the drop-down list and click the Add button.

Finally, we need to declare support for the cable. Click on the *SerialConsole-Info.plist* file to open it in the editor. Right-click on the bottommost row of the list and select Add Row from the menu. An extra row will be added to the table, and you'll be presented with a drop-down menu. Type **UISupportedExternalAccessoryProtocols** into the box. This will change to the human-readable text "Supported external accessory protocols." Type the string **com.redpark.hobdb9** into Item 0 to declare our support for the cable.

The CorePlot Library

Unsurprisingly perhaps when dealing with sensor data, the need to display a graph or a plot of some kind comes up fairly often when you're developing applications.

Unfortunately, there is no native support for graphs in the iOS SDK. The Core Plot library fills that hole (see Figure 4-3).

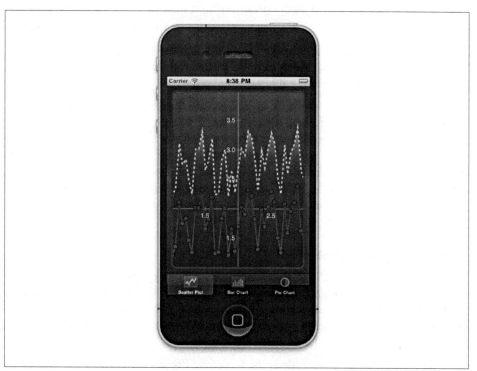

Figure 4-3. The CorePlot test application running in the iPhone Simulator

The Core Plot library is a third-party 2D plotting framework for iOS and Mac OS X, tightly integrated with Core Animation and under active development. At the time of this writing, the latest release was version 0.4. You can download it from *http://code .google.com/p/core-plot/*.

If you want to live on the bleeding edge, you can check out the source code straight from Core Plot's repository using Mercurial:

```
hg clone http://core-plot.googlecode.com/hg/ core-plot
```

Mercurial doesn't ship with Xcode, so unless you've already installed it, your first step before doing that will probably be to install Mercurial itself. You can download the latest version of Mercurial from *http:// mercurial.selenic.com/wiki/*.

Compiling the Core Plot library from source code

Download the *CorePlot_0.4.zip* file from the Core Plot website, or check the latest version out of the project's Mercurial repository. Open up the *CorePlot/framework/* folder and click on the *CorePlot-CocoaTouch.xcodeproj* project file to open it in Xcode.

Select "Universal Library | iOSDevice" from the Scheme drop-down menu and then Build (⌘B) from the Product menu to build a universal (fat) static library that we can use on both the device and in the iPhone Simulator.

After building, move one folder back up and go into the newly created *build* directory. Inside the *build/Release-universal/* directory you'll find the *libCorePlot-CocoaTouch.a* static library; however, in addition to this, we need the associated header files. The custom build script, which created the universal library, doesn't copy these header files from the individual build directories. However you can find them in the *build/Release-iphoneos/usr/local/include/* or *build/Release-iphonesimulator/usr/local/include/* directories. Which set of header files you use in your project isn't really relevant; both sets are identical.

Adding the Core Plot library to the project

Drag and drop the *libCorePlot-CocoaTouch.a* static library from the *build/Release-universal/* folder into your project, remembering to tick the "Copy items into destination group's folder (if needed)" checkbox when prompted. Then copy one of the two sets of header files, all of them, into your project. Again remember to tick the "Copy items into destination group's folder (if needed)" checkbox.

Click on the MaxSonar project file at the top of the Project pane, select the MaxSonar Target, and click on the Build Settings tab. Find the Other Linker Flags setting (under the Linking section) and double-click to open the pop-up window and add the `-ObjC -all_load` flags to the target (see Figure 4-4).

Finally, since the Core Plot library makes use of the QuartzCore framework, we need to add this to our project. Click on the Build Phases tab and then on the Link Binary with Libraries item to open up the list of linked frameworks, and click on the + symbol to add a new framework. Select the QuartzCore framework from the drop-down list and click the Add button.

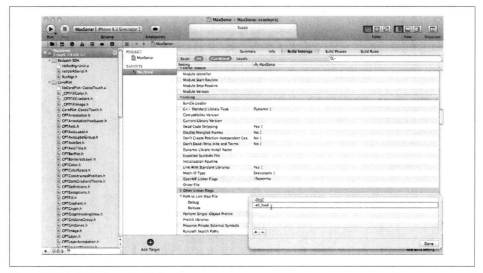

Figure 4-4. Adding the other linker flags to your target

Building the User Interface

Now that we've added the third-party infrastructure we'll need to build the application, it's time to build out the user interface. Click on the *MaxSonarViewController.xib* file to open it in Interface Builder and drag and drop a `UINavigationBar`, a `UISwitch`, a couple of `UILabel` elements, and finally a generic `UIView` into your View. Arrange them as shown in Figure 4-5. Change the text in the navigation bar to read "LV-MaxSonar-EZ1" and the text in the two labels to be "Distance:" and "0.00".

Click on the generic `UIView` in your View and open up the Identity inspector in the Utilities pane on the right. We need to change the Class of our View from `UIView` to be a `CPTGraphHostingView`. See Figure 4-6.

> Depending on which version of the Core Plot library you are working with, you may need to specify the custom view as a `CPGraphHosting View` instead.

This custom view is the one we're going to use to plot our real-time graph of the sensor readings coming from the EZ1 sensor via the Arduino and the serial link to our iPhone.

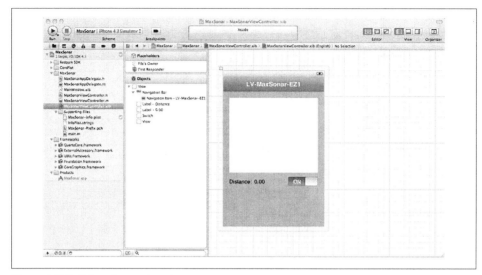

Figure 4-5. Building the user interface

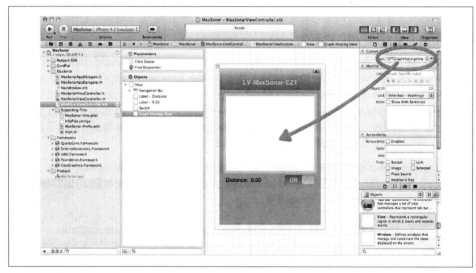

Figure 4-6. Setting a custom view

After changing the view's class, close the Utilities panel and open up the Assistant Editor. Make sure it's in Automatic mode and Ctrl-click and drag from the CPTGraph HostingView to the *MaxSonarViewController.h* interface file to create a graph instance variable and property. Similarly, Ctrl-click and drag from the right-hand UILabel and UISwitch to create distance and toggle instance variables and properties. Finally, Ctrl-click and drag one more time from the UISwitch to a toggled method and IBAction, as in Figure 4-7.

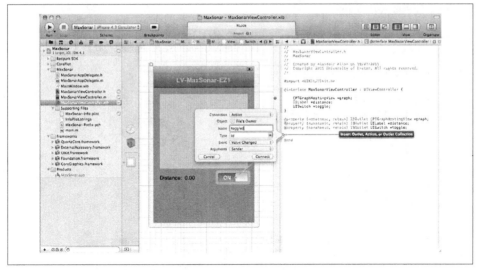

Figure 4-7. Connecting the outlets and actions

Save your changes, switch back to the Standard Editor, and click on the *MaxSonar-ViewController.h* interface file to open it in the editor. You should notice that we've got an error next to the `CPTGraphHostingView` instance variable; that's because Xcode doesn't know what it is yet. Go ahead and add the following import statement at the top of your code:

```
#import "CorePlot-CocoaTouch.h"
```

If you select Build (⌘B) from the Product menu to build your project at this point, everything should compile. If that's not the case, you should probably make sure you've added the External Accessory and QuartzCore frameworks to your project before you begin looking elsewhere.

Building the Backend

We're done with the user interface. What we need to do now is connect our interface to the incoming measurements from our Arduino board. Obviously the first thing we need to do is integrate the Redpark library into our code. The code for this is going to look almost identical to the code we've seen in the last couple of chapters. However, we're also going to have to establish some infrastructure for our plot. We'll use a simple XY scatter plot for that task.

Open the *MaxSonarViewController.h* interface file and add the following code:

```
#import <UIKit/UIKit.h>
#import "CorePlot-CocoaTouch.h"
#import "RscMgr.h"

#define BUFFER_LEN 1024
```

```
#define MAX_POINTS 200

@interface MaxSonarViewController : UIViewController
                            <CPTPlotDataSource, RscMgrDelegate> {

    CPTGraphHostingView *graph;

    CPTXYGraph *plot;
    NSMutableArray *dataForPlot;

    UILabel *distance;
    UISwitch *toggle;

    RscMgr *manager;
    UInt8 rxBuffer[BUFFER_LEN];
    UInt8 txBuffer[BUFFER_LEN];
}

@property (nonatomic, retain) IBOutlet CPTGraphHostingView *graph;
@property (nonatomic, retain) IBOutlet UILabel *distance;
@property (nonatomic, retain) IBOutlet UISwitch *toggle;
@property (nonatomic, retain) NSMutableArray *dataForPlot;

- (IBAction)toggled:(id)sender;

- (NSNumber *) yValueForIndex:(NSUInteger)index;

@end
```

This makes our class both a Core Plot data source and a Redpark Serial Cable delegate object. Now open the corresponding *MaxSonarViewController.m* implementation file. We need to remember to synthesize our new property:

```
@synthesize dataForPlot;
```

and then release it in the `dealloc` method:

```
    [dataForPlot release];
```

and finally set it to nil in `viewDidUnload` method:

```
    [self setDataForPlot:nil];
```

We want to go ahead and add a `viewDidLoad` method. Here we're going to set up the cable manager object as usual, but also our Core Plot plot object. Doing that setup involves a lot of boilerplate code, and I'm not really going to talk about it in detail here, as it's all a bit tangential to what we're actually doing:

```
- (void)viewDidLoad {
    [super viewDidLoad];

    // cable manager
    rscMgr = [[RscMgr alloc] init];
    [rscMgr setDelegate:self];

    // Create plot from theme
```

```
plot = [[CPTXYGraph alloc] initWithFrame:CGRectZero];
CPTTheme *theme = [CPTTheme themeNamed:kCPTPlainBlackTheme];
[plot applyTheme:theme];
graph.collapsesLayers = NO;
graph.hostedGraph = plot;

plot.paddingLeft = 10.0;
plot.paddingTop = 10.0;
plot.paddingRight = 10.0;
plot.paddingBottom = 10.0;

// Setup plot space
CPTXYPlotSpace *plotSpace = (CPTXYPlotSpace *)plot.defaultPlotSpace;
plotSpace.allowsUserInteraction = YES;
plotSpace.xRange = [CPTPlotRange plotRangeWithLocation:CPTDecimalFromFloat(1.0)
                              length:CPTDecimalFromFloat(MAX_POINTS)];
plotSpace.yRange = [CPTPlotRange plotRangeWithLocation:CPTDecimalFromFloat(1.0)
                              length:CPTDecimalFromFloat(MAX_POINTS)];

// Axes
CPTXYAxisSet *axisSet = (CPTXYAxisSet *)plot.axisSet;

CPTXYAxis *x = axisSet.xAxis;
x.majorIntervalLength = CPTDecimalFromString(@"100");
x.minorTicksPerInterval = 4;
x.minorTickLength = 5.0f;
x.majorTickLength = 7.0f;

CPTXYAxis *y = axisSet.yAxis;
y.majorIntervalLength = CPTDecimalFromString(@"50");
axisSet.yAxis.minorTicksPerInterval = 2;
axisSet.yAxis.minorTickLength = 5.0f;
axisSet.yAxis.majorTickLength = 7.0f;

// Create a green plot area
CPTScatterPlot *line = [[[CPTScatterPlot alloc] init] autorelease];
CPTMutableLineStyle *lineStyle = [CPTMutableLineStyle lineStyle];
lineStyle.miterLimit = 1.0f;
lineStyle.lineWidth = 3.0f;
lineStyle.lineColor = [CPTColor greenColor];
line.dataLineStyle = lineStyle;
line.identifier = @"Green Plot";
line.dataSource = self;
[plot addPlot:line];

// Do a green gradient
CPTColor *areaColor = [CPTColor colorWithComponentRed:0.3
                              green:1.0 blue:0.3 alpha:0.8];
CPTGradient *areaGradient = [CPTGradient gradientWithBeginningColor:areaColor
                              endingColor:[CPTColor clearColor]];
areaGradient.angle = -90.0f;
CPTFill *areaGradientFill = [CPTFill fillWithGradient:areaGradient];
line.areaFill = areaGradientFill;
line.areaBaseValue = [[NSDecimalNumber zero] decimalValue];
```

```
// Add plot symbols
CPTMutableLineStyle *symbolLineStyle = [CPTMutableLineStyle lineStyle];
symbolLineStyle.lineColor = [CPTColor blackColor];
CPTPlotSymbol *plotSymbol = [CPTPlotSymbol ellipsePlotSymbol];
plotSymbol.fill = [CPTFill fillWithColor:[CPTColor greenColor]];
plotSymbol.lineStyle = symbolLineStyle;
plotSymbol.size = CGSizeMake(10.0, 10.0);
line.plotSymbol = plotSymbol;

CABasicAnimation *fadeInAnimation = [CABasicAnimation
                                     animationWithKeyPath:@"opacity"];
fadeInAnimation.duration = 1.0f;
fadeInAnimation.removedOnCompletion = NO;
fadeInAnimation.fillMode = kCAFillModeForwards;
fadeInAnimation.toValue = [NSNumber numberWithFloat:1.0];

// Add some initial data
NSMutableArray *contentArray = [NSMutableArray arrayWithCapacity:200];
self.dataForPlot = contentArray;
}
```

Now we need to add the CorePlot `CPTPlotDataSource` data source methods:

```
#pragma mark - CPTPlotDataSource Methods

-(void)pointReceived:(NSNumber *)point{

    CPTPlot *thisPlot = [plot plotWithIdentifier:@"Green Plot"];
    [self.dataForPlot addObject:point];
    [thisPlot insertDataAtIndex:self.dataForPlot.count-1 numberOfRecords:1];
    self.distance.text = [NSString stringWithFormat:@"%@", point];

}

-(NSUInteger)numberOfRecordsForPlot:(CPTPlot *)plot {
    return [dataForPlot count];
}

-(NSNumber *)numberForPlot:(CPTPlot *)plot field:(NSUInteger)fieldEnum recordIndex:
(NSUInteger)index {

    if (fieldEnum == CPTScatterPlotFieldX) {
        return [NSNumber numberWithInteger:index];
    } else {
        return [self yValueForIndex:index];
    }
}
```

Along with our `yValueForIndex:` convenience method:

```
- (NSNumber *) yValueForIndex:(NSUInteger)index {
    return [self.dataForPlot objectAtIndex:index];
}
```

Finally add the Redpark Serial delegate methods:

```
#pragma mark - RscMgrDelegate methods

- (void) cableConnected:(NSString *)protocol {
    [rscMgr setBaud:9600];
    [rscMgr open];

}

- (void) cableDisconnected {
}


- (void) portStatusChanged {
}

- (void) readBytesAvailable:(UInt32)numBytes {
    int bytesRead = [rscMgr read:rxBuffer Length:numBytes];

    NSString *string = nil;
    for(int i = 0;i < numBytes;++i) {
        if ( string ) {
            string = [NSString stringWithFormat:@"%@%c", string, rxBuffer[i]];
        } else {
            string = [NSString stringWithFormat:@"%c", rxBuffer[i]];
        }
    }
    [self pointReceived:[NSNumber numberWithInt:[string intValue]]];

}

- (BOOL) rscMessageReceived:(UInt8 *)msg TotalLength:(int)len {
    return FALSE;
}

- (void) didReceivePortConfig {
}
```

Check that everything builds and runs by running the app in the simulator. You should see something like Figure 4-8. If that looks good, switch to the device and build and deploy the application onto your iPhone.

If everything has gone well, at this point we need to put the iPhone application, along with Xcode itself, to one side for now. It's time to look at the Arduino end of our build.

Writing the Arduino Sketch

Open up the Arduino development environment and enter the following code and upload it to your Arduino board. We'll be using analog pin 0, normally marked as A0 on most boards, for the analog input from the EZ1 sensor:

Figure 4-8. The MaxSonar app running in the iPhone Simulator

```
int statusLed = 13;
int ez1Analog = 0;

void setup() {
   pinMode(statusLed,OUTPUT);
   pinMode(ez1Analog,INPUT);
   Serial.begin(9600);
}

void loop() {
   int val = analogRead(ez1Analog);
   if (val > 0) {
      val = val / 2;
      float cm = float(val)*2.54;
      Serial.println( int(cm) );
   }
   blinkLed( statusLed, 100 );①
}

void blinkLed(int pin, int ms) {
   digitalWrite(pin,LOW);
   digitalWrite(pin,HIGH);
   delay(ms);
   digitalWrite(pin,LOW);
   delay(ms);
}
```

❶ We'll use a convenience method here to blink the board LED on pin 13 for 100 ms. First, so we know something is actually happening, but also to space out the readings being sent to the phone.

Putting It All Together

Nominally the EZ1 requires 5 V; however, according to the datasheet, we can operate it at 3.3 V, albeit with somewhat reduced performance. This simplifies connecting all of our components to the Arduino board. Plug the +VE wire from the EZ1 into the 3.3V pin on the Arduino board, and the GND wire into one of the available GND pins, then connect the AN wire to analog pin 0 (A0). Finally connect your RS-232 to TTL serial adaptor to the 5V, GND, RX, and TX pins as normal (see Figure 4-9).

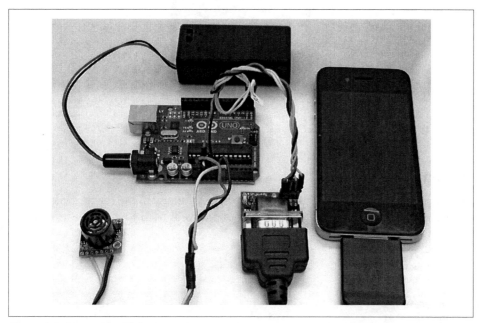

Figure 4-9. Connecting all the components

If you do want to connect the EZ1 to the +5V pin rather than the 3.3V pin, probably the easiest thing to do would be to use a breadboard and some jumper wires.

Once you're satisfied that everything is connected correctly, power on the Arduino board and start up the application. Point the EZ1 toward the ceiling to get a decent distance measurement, and then place your hand over the sensor. You should see the measured distance drop dramatically. Slowly raise your hand, holding it above the sensor to see the reading change. Hopefully you'll see something that looks a lot like Figure 4-10 at this point.

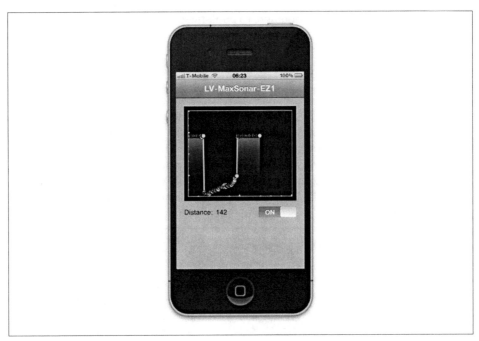

Figure 4-10. The MaxSonar application running on the iPhone

Turning Things On and Off

You've probably noticed that we haven't done anything with the toggle switch. Let's go ahead and make that usable now. First of all, we'll need to change our Arduino sketch:

```
int statusLed = 13;
int powerLed = 12;
int ez1Analog = 0;

int aByte;
int flag = 0;

void setup() {
   pinMode(statusLed,OUTPUT);
   pinMode(ez1Analog,INPUT);
   Serial.begin(9600);

   pinMode(powerLed,OUTPUT);
   blinkLed( statusLed, 500 );
}

void loop() {

   if (Serial.available() > 0) {
        aByte = Serial.read();
```

```
            if ( flag == 0 ) {
               flag = 1;
               digitalWrite(powerLed, HIGH);
            } else {
               flag = 0;
               digitalWrite(powerLed, LOW);
            }
        }

     if ( flag == 1 ) {
       int val = analogRead(ez1Analog);

       if ( val > 0 ) {
            val = val / 2;
            float cm = float(val)*2.54;
            Serial.println( int(cm) ); // cm
            blinkLed( statusLed, 100);
        }
     }

  }

  void blinkLed(int pin, int ms) {
     digitalWrite(pin,LOW);
     digitalWrite(pin,HIGH);
     delay(ms);
     digitalWrite(pin,LOW);
     delay(ms);
  }
```

Here we've rewritten our loop( ) so that the Arduino only sends data to the iPhone if it has been toggled on, by receiving a byte from the phone, and then stops transmitting data if it receives another byte. I'm also using a second LED in pin 12 to indicate that data is being sent, but that's purely optional, and you can omit it if you can't be bothered to wire another LED up to your Arduino board.

You'll notice that we're assuming an initial state of "off," but that currently our user interface has an initial state of "on." We need to fix that, so open up the *MaxSonar-ViewController.xib* file in Interface Build, click on the switch in the view, and in the Attributes inspector in the Utilities panel, change the initial switch state to "off."

Finally, to glue everything together, we add some code to the toggled callback:

```
  - (IBAction)toggled:(id)sender {
     txBuffer[0] = 1;
     int bytesWritten = [rscMgr write:txBuffer Length:1];

  }
```

Now if you rebuild and then redeploy the MaxSonar application to the iPhone, you should be able to stop and start the data coming from the phone using the switch.

Connecting Directly to the Cable

Because the EZ1 has an RS-232 connection, it's actually possible to connect it directly to the Redpark cable without using the Arduino board at all. The Redpark Serial Cable is wired in a standard fashion, terminating with a male DB-9 connector; the pin out for this connector is shown in Figure 4-11 and described in Table 4-1.

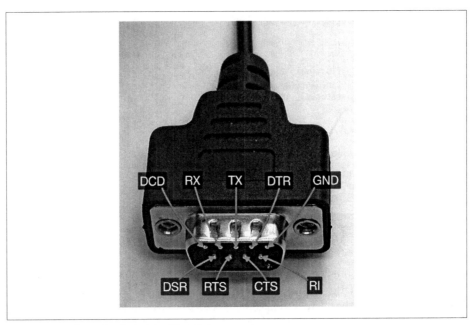

Figure 4-11. The Redpark Serial cable DB-9 connector

Table 4-1. The Redpark Serial Cable DB-9 (male) Connector Pin Out

Pin Number	Function	I/O	Description
1	DCD	I	Data carrier detect
2	RX	I	Receive
3	TX	O	Transmit
4	DTR	O	Data terminal ready
5	GND	-	Ground
6	DSR	I	Data set ready
7	RTS	O	Request to send
8	CTS	I	Clear to send
9	RI	I	Ring indicator

Comparing Figure 4-1 and Figure 4-11, we'll need to wire up only 3 of the 9 available pins: GND, RX, and TX. Taking a female DB-9 connector, you should therefore wire it as in Figure 4-12. Comparing Figure 4-11 and Figure 4-12, you should notice that in the female connector, we wire pin 2 to TX and pin 3 to RX. This means that TX from the EZ1 is connected to RX on the cable, and vice versa.

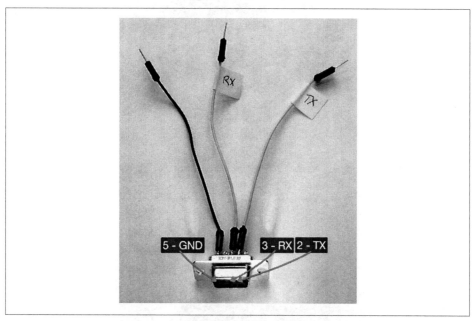

Figure 4-12. Pin wiring for the LV-MaxSonar-EZ1 connecting to DB9 (female)

After wiring this cable, go ahead and solder a header block onto the EZ1 sensor board (see Figure 4-13).

When wiring the EZ1 sensor, you should be careful not to obstruct the sensor head itself; doing so means that you'll get inaccurate measurements when you power it on. Depending on your application, you may want to attach the header pins to the rear of the board or use a right angle header block so that there is no chance of obstructing the sensor.

Once you've soldered the header block to the sensor board, you should find an appropriate 3.3 V or 5 V power supply. I'm using 2 × 1.5 V batteries to provide a 3 V supply, which should be sufficient to test the sensor. We then need to gather our ground connections and power the sensor. The easiest way to do this is probably to use a breadboard (see Figure 4-14).

Figure 4-13. Header block soldered to the EZ1 sensor

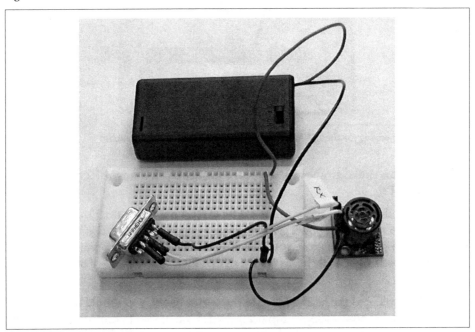

Figure 4-14. Wiring the EZ1 sensor, cable, and battery pack

Now that the hardware is complete, let's look and see what we need to do on the software side. Looking at the datasheet, we see that the output from the sensors is an ASCII capital "R", followed by three ASCII character digits representing the range in inches (up to a maximum of 255), followed by a carriage return (ASCII character 13), for a message length of 5 bytes in total.

We've already written some software that will handle that sort of output just fine, our SerialConsole application from Chapter 2. Connect your Redpark cable to the female DB9 connector and to your iPhone and run the SerialConsole application. You should see something like Figure 4-15.

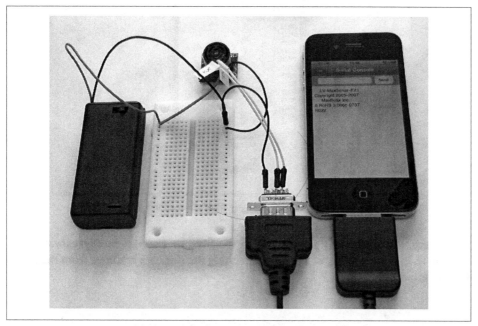

Figure 4-15. The output from the EZ1 inside the SerialConsole application

You'll see that the datasheet isn't strictly accurate; a copyright notice is displayed the first time the sensor is powered on. Good thing we checked. If you type something in the text entry widget and tap on send, you'll see another reading appears in the console view (see Figure 4-16).

Figure 4-16. The SerialConsole application showing multiple readings from the EZ1

From here, it's a relatively simple job to modify our existing `readBytesAvailable:` method to look for a string of the form matching the distance readings from the EZ1's RS-232 interface.

CHAPTER 5

Connecting to an XBee Network

All of the incarnations of the iPhone, iPod touch, and iPad come with WiFi networking, or more formally, IEEE 802.11. This is a set of standards for implementing wireless local area networking (WLAN) computer communication in the 2.4, 3.6, and 5 GHz frequency bands, although none of the existing iOS-based devices support anything other than the 2.4 GHz band.

However, there are many other standards for wireless communication, many of which also make use of the 2.4 GHz band, which are intended for other purposes and hence designed in a different fashion. They may use the same wavelength bands, but the protocols are very different.

One of the most commonly used protocols when dealing with sensor networks is IEEE 802.15.4. In contrast to IEEE 802.11, this protocol focuses on low-cost, low-speed communication between devices and is intended for low-power scenarios. It is the basis for the higher-level ZigBee protocol, also commonly found when dealing with sensor networks, which further extends the standard by developing the upper layer protocols to enable mesh networking between devices.

While many other bands of 802.15.4-based hardware exist, probably the most heavily used are the XBee branded radios. They're in common use, both by professionals who use them for prototyping, and in the hobbyist market, because they're especially easy for beginners to use. For this and other reasons I'm going to focus on the XBee-branded radio modules that are manufactured by Digi International (*http://www.digi.com/*).

XBee Modules

Digi manufactures a confusingly large array of XBee-branded radios; a quick count shows that there are at least 30 different combinations of hardware, firmware, and antenna options, and this number is only going to get larger. Unfortunately, it's very easy to get confused by the growing range of XBee-branded radio modules.

There are two types of pin-compatible XBee modules. The Series 1 module is intended for point-to-point and point-to-multipoint applications, while the Series 2 module is intended for applications that require mesh networking. In this chapter I'll be using the Series 1 radio modules (see Figure 5-1).

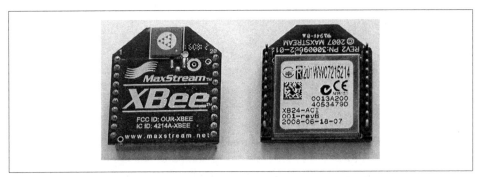

Figure 5-1. An XBee Series 1 module with a chip antenna

I use Series 1 radios mainly because they're very easy for beginners to use and to configure and, perhaps more importantly, easy to obtain. You can get Series 1 radios from a number of vendors, including SparkFun, which carries a wide range of modules (see *http://www.sparkfun.com/categories/111*), although, despite the dizzying range on display, they still don't carry them all. You can also get the radios from Maker Shed (*http://www.makershed.com/ProductDetails.asp?ProductCode=MKAD14*).

Series 1 or Series 2?

One of the misunderstandings when dealing with XBee modules is that somehow Series 2 modules are "better" than Series 1 modules. That's just not true. Instead, the two types are intended for different purposes. The Series 1 modules are popular with the hobbyist market, and they are an excellent fit for a straightforward cable replacement or for small networks. Series 2 implements the full ZigBee protocol, and the modules are intended for larger networks.

 While the XBee Series 1 and XBee Series 2 modules have the same form factor and are pin-for-pin compatible, they are based on different chip-sets and are running different protocols, so they are not over-the-air compatible. You cannot mix Series 1 and Series 2 modules in the same wireless network.

A Series 1 module will work out of the box for point-to-point communication (i.e., cable replacement) without any work, which is what we're going to be using them for in this chapter. So if you're just starting out with wireless networks, I'd recommend you start with the Series 1 radios.

If you're interested in building larger XBee-based networks, you might want to look at Building Wireless Sensor Networks (*http://oreilly.com/catalog/9780596807740*) by Robert Faludi (O'Reilly). While that book uses Series 2 modules exclusively, it can also help you with Series 1 networks, as Series 1 commands are, for the most part, just a subset of the Series 2 command set. There are also several interesting projects that make use of the Series 1 modules I talk about in this chapter in Making Things Talk (*http://oreilly.com/catalog/0636920010920*) by Tom Igoe (O'Reilly).

Regular Versus Pro?

Both Series 1 and Series 2 modules come in two flavors: regular and Pro. The Pro versions of the modules are slightly larger, although despite this they are still pin-compatible, and they use more power. Crucially, however, they have a much longer range. Predictably, they also cost a lot more, so unless you need the extended range they offer, you should stick with the regular XBee modules. Since they're pin-compatible, you can always drop a Pro into your build later on if you decide you need the extended range (and can afford the increased power load).

802.15.4 or ZigBee?

There is a lot of confusion surrounding the two XBee protocols. Series 1 modules use the IEEE 802.15.4 standard protocol, which allows point-to-point and point-to-multipoint networking, although they also have a proprietary mesh networking protocol. Series 2 modules use the ZigBee protocol, which is a mesh networking standard, built on top of the 802.15.4 protocol, which is spoken by numerous other pieces of hardware, not just the XBee radio modules.

Which Aerial?

There are four different types of antenna on offer: chip, wire, UFL, and RP-SMA. The chip and wire antennas come preconnected to the radio module; however, the UFL and RP-SMA offerings ship with connectors only on the board. You'll need to purchase an appropriate aerial with the proper connector to get them to function. Unless you're intending to enclose your project in a box, at which point mounting the antenna outside the box using a UFL or RP-SMA aerial is probably a good idea, or need the high-power transmission they provide, you should be safe enough to stick with chip- or wire-based aerials.

How to Configure an XBee Series 1 Radio

To configure an XBee module, we need to be able to talk to it. Generally, we'll be using your Mac to configure your XBee, although there is nothing stopping us from writing an iPhone or iPad application to allow us to do this configuration. For now, we'll need a way to connect the XBee to your Mac's USB port.

Since they were designed to be soldered directly to a PCB, XBee modules rather unfortunately use 2 mm headers instead of the more familiar 0.1 in headers we are all so used to. They therefore generally require a breakout board of some kind before they can be used on a standard bread- or proto-board, and an adaptor before they can be connected to your Mac.

There are a number of breakout boards on the market. Generally, you'll just need a simple board that breaks the XBee pins out to standard 0.1 in header blocks, like SparkFun's Breakout Board for XBee Module (*http://www.sparkfun.com/products/ 8276*).

To connect the XBee to your computer, you'll need a more complicated board that takes care of regulating the input voltage to the 3.3 V required by the module, signal conditioning, and providing basic activity indicator LEDs. Again, there are a large number of these on the market, like SparkFun's XBee Explorer Regulated (*http://www .sparkfun.com/products/9132*) and Adafruit's XBee Adaptor Kit (*http://www.adafruit .com/products/126*). See Figure 5-2.

There are even boards allowing you to plug the XBee directly into your Mac's USB port, such as SparkFun's XBee Explorer Dongle (*http://www.sparkfun.com/products/9819*); see Figure 5-2 again.

Almost invariably, XBee USB adaptors require drivers from FTDI (*http: //www.ftdichip.com/Drivers/VCP.htm*). You need to make sure you have installed these before using your adaptor. If you have an older Arduino board, you may have already installed the correct driver as boards before the Uno made use of the same USB chipset.

Figure 5-2. Adafruit XBee Adaptor (left) and the SparkFun XBee Explorer (right)

Connecting the XBee to Your Mac

Because I already had a couple lying around, I'm going to use the Adafruit board, which is also carried in the Maker Shed (*http://www.makershed.com/ProductDetails.asp?Pro ductCode=MKAD13*). It's designed to be used with an FTDI cable (see Figure 5-3), and because of that, it can also be easily used with a breadboard to connect the XBee to an Arduino, so I won't need a separate breakout board.

Figure 5-3. XBee module mounted on an Adafruit Adaptor Kit connected to an FTDI cable for programming

While Digi's own configuration tool, X-CTU, is available for free, it's also Windows-only. Unless you're running Windows under BootCamp or in a virtual machine, you won't be able to use it. Luckily, it's only needed infrequently to update the firmware of the XBee modules. For configuring the radios, we can generally just use a terminal program.

There are many terminal programs available for Mac OS X. I'm going to use CoolTerm by Roger Meier (*http://freeware.the-meiers.org/*). This is a relatively simple application that's perfectly suited to allow you configure your XBee radios.

Plug your XBee radio module into your adaptor board, and connect the board to your Mac. If you're using the Adafruit board as I am, you should see the green (ASC) LED on the board start to flash.

Open up your terminal application and make sure it's set to the correct serial port. You can find out which serial ports are available by using the command line:

```
% ls /dev/tty.*
crw-rw-rw-  1 root  wheel  11,   2  8 Jul 09:01 /dev/tty.Bluetooth-Modem
crw-rw-rw-  1 root  wheel  11,   0  8 Jul 09:01 /dev/tty.Bluetooth-PDA-Sync
crw-rw-rw-  1 root  wheel  11,  66  3 Aug 14:15 /dev/tty.usbserial-FTE4XVKD
```

The correct port for the XBee module should (at least normally) be fairly obvious. In my case, it is /dev/tty.usbserial-FTE4XVKD. If you are using CoolTerm, you can change the attached serial port by clicking on the Options button in the toolbar (see Figure 5-4).

> By default, CoolTerm has Local Echo turned off. You may want to enable this in the Options menu to make things easier on yourself. If it's turned off, you won't see what you're typing, and only the responses from the XBee radio will be displayed.

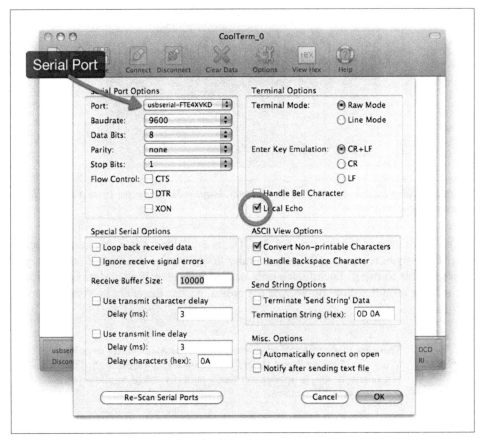

Figure 5-4. Configuring the port and local echo in CoolTerm options

If your terminal program doesn't default to it, set it to use 9,600 baud and 8-N-1 with no flow control.

Once you're sure you have your terminal program set up correctly, connect it to the selected serial port. In CoolTerm you do this by clicking on the Connect button in the toolbar. At this point, type +++ (three plus signs in fairly quick succession) into the main window. If everything is connected correctly, you should get an **OK** back from the XBee radio module (see Figure 5-5).

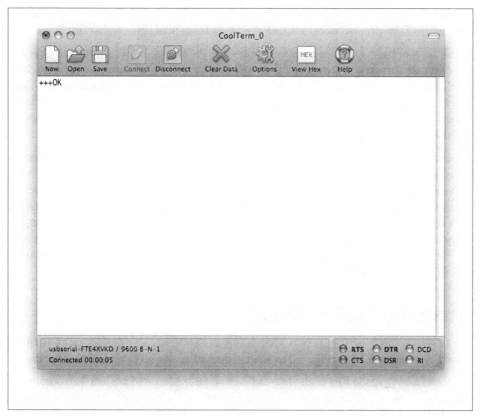

Figure 5-5. The XBee radio module is correctly connected to your Mac

By typing the three plus symbols, we have put the XBee into configuration mode. At this point we can send **AT** commands to the module, which will allow us to configure the radio. After a while, the XBee will time out and go back to pass-through connection mode. If that happens while you're configuring it, just type in **+++** and it will start responding once again.

The basic configuration on the XBee consists of the baud rate for serial communication, a network identifier (PAN), the node address (MY), and the destination node address (DL).

XBee Addressing

The 802.15.4 protocol uses addressing to distinguish one radio from the next, and to prevent duplicate packets. It is very important that each module have a unique source address (MY) to prevent nonduplicate messages from being ignored as duplicates.

There are two basic forms of addressing between the modules: Broadcast and Unicast. A Broadcast message is a message that will be received by all modules on any given PAN. The message is sent only once and not repeated, so there is no guarantee of any given node receiving the message. In order to send a Broadcast message, set the destination node address (DL) to 0xFFFF. With these settings, all XBee modules within range of the broadcasting node will receive the message.

A Unicast message is more reliable. It is sent from one module to any other module based on the module's addressing. If the message is properly received, the receiving radio will send back an acknowledgment (or ACK). If the transmitting module doesn't receive an ACK, it will attempt MAC level retries, three retries for every transmission, for a total of four attempts, until it receives an ACK. This greatly increases the probability of getting the data through to the destination. In order to send a Unicast message, set the destination node address (DL) to be the node address (MY) of the node with which you wish to communicate.

Configuring Two XBee Radios

We're now going to configure two XBee radios as a straight cable replacement. In other words, the first will be configured to send messages to the second, and vice versa.

Go ahead and plug the first of the two radios into your Mac, and place it into command mode, as in the previous section, by sending a **+++** to the appropriate serial port from your terminal program.

The first thing we need to do is set the baud rate. When in configuration mode, enter **ATBD**. You should get a number between 0 and 7 returned, as below:

```
ATBD
3
```

This number indicates that the radio is set to 9,600 baud. We'll leave our radio at 9,600 baud for now. However, if you wanted to set the radio to a different baud rate, you'd issue the command **ATDB***n*, where *n* was a value corresponding to the baud rate you wanted from Table 5-1.

Table 5-1. The value returned by the ATBD command and corresponding baud rate

ATBD Value	Baud Rate
0	1200
1	2400
2	4800
3	9600
4	19200
5	38400
6	57600
7	115200

So, for example, to set our radio to 57,600 baud, we'd issue the command **ATDB6**:

```
ATBD6
OK
```

We can then check the baud rate has been set, by again issuing the **ATBD** command. This time it should return a six:

```
ATBD
6
```

Having set the baud rate, we need to go ahead and similarly set the PAN ID using the **ATID** command, the node address using the **ATMY** command, and the destination node address using the **ATDL** command. Finally, we need to commit our changes by issuing a write command, **ATWR**, to save our settings.

Therefore, for the first of our two radios, we should issue the following commands:

```
+++
ATBD 3
ATID 1111
ATMY 2345
ATDL 7890
ATWR
```

You should get an OK back from the modem in all cases. You may want to check the values along the way; see, for instance, Figure 5-6.

Once you've configured the first radio, disconnect the serial port and unplug the XBee module from the adaptor. Then plug the second module into the adaptor, connect to the appropriate serial port, and issue the following commands:

```
+++
ATBD 3
ATID 1111
ATMY 7890
ATDL 2345
ATWR
```

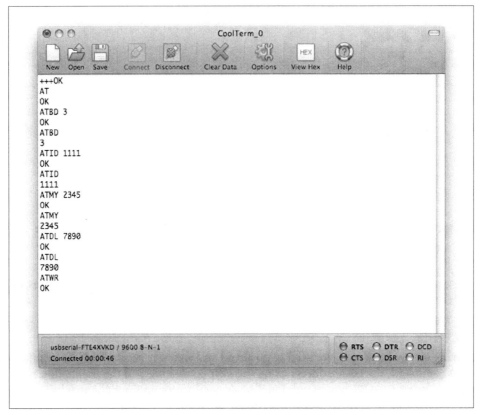

Figure 5-6. Configuring the first of the two XBee modules

This configures the second radio as the destination of the first, and vice versa. See Table 5-2.

Table 5-2. Radio configuration

Radio	PAN	MY	DL
1	1111	2345	7890
2	1111	7890	2345

Unplug the second radio once you've finished configuring it. We'll get back to the XBee modules later. However, now that we've configured them, we're going to go ahead and test out the connection by attaching one of them to an Arduino and the other to our Mac.

Connecting an XBee to an Arduino

You can actually directly connect an XBee radio module to an Arduino by making use of the 3.3V pin on the Arduino board to provide a regulated voltage for the module. See Figure 5-7.

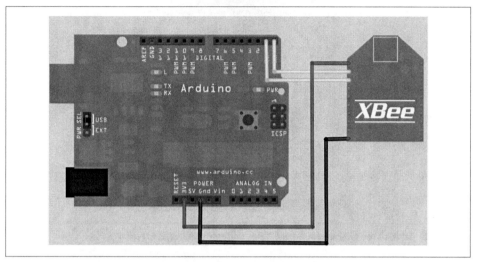

Figure 5-7. Connecting an XBee to an Arduino

However, it's a lot easier if you make use of a breakout board, or an appropriate adaptor board, like the Adafruit board we used to connect the XBee to our Mac earlier (see Figure 5-8).

Alternatively, if you're going to be making a lot of use of XBee modules, you might want to invest in an Arduino Fio board (see Figure 5-9), which is an Arduino board with an attached XBee socket. You can buy an Arduino Fio from SparkFun (*http://www .sparkfun.com/products/9712*).

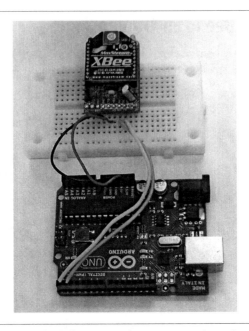

Figure 5-8. An Arduino Uno and Adafruit XBee adaptor connected together

Figure 5-9. The Arduino Fio with XBee connected to a LiPo battery

We're going to use the same simple sketch on our Arduino that we used in Chapter 1 to print "Hello World" to the serial port. Connect your Arduino board to your Mac and upload the following sketch:

```
void setup() {
  Serial.begin(9600);
}

void loop() {
  while (Serial.available() <= 0) {
```

```
        Serial.println("Hello world");
        delay(300);
    }
}
```

 You cannot upload a sketch from your Mac to the board with the XBee, or any other hardware, connected to pins 0 (RX) and 1 (TX), as this interferes with the USB serial connection from your Mac. You therefore must detach the XBee from your Arduino board before uploading your sketch.

Disconnect your Arduino board from your Mac and connect it to an external power supply.

Once you've done that, attach one of the XBee radios to your Mac and open up the CoolTerm application once again. Connect to the appropriate serial port so that you can see the output from the radio.

Then go ahead and connect the other radio to your Arduino board, as in Figure 5-8, and power it on. You should see something very much like Figure 5-10 in your terminal program.

Figure 5-10. The Arduino says "Hello World" over an XBee connection

Connecting an XBee to an iOS Device

Now that we've tested things between our Arduino board and our Mac, what do we need to do the same thing between our Arduino board and our iPhone?

XBee to RS-232 Serial

The first thing we need is an RS-232 adaptor. Fortunately SparkFun sells the XBee Explorer Serial (*http://www.sparkfun.com/products/9111*) for $29.95 (see Figure 5-11).

I don't know of any others on the market that are as easily available, although if you don't want to purchase the SparkFun adaptor, it should be moderately easy to bread-board an adaptor using your RS-232 to TTL serial adaptor that we've been using up till this point.

However, if you do that, you should remember that the XBee module runs at 3.3 V and the Redpark cable is at 5 V. You'll need a voltage level shifter between the 3.3 V RX and TX of the XBee module and the 5 V of the TTL serial adaptor.

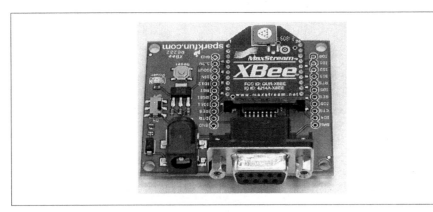

Figure 5-11. The SparkFun XBee Explorer Serial

Go ahead and unplug the XBee that was originally attached to your Mac and plug it into the serial adaptor. Then plug the serial adaptor into your Redpark cable, and that in turn into your iPhone. Attach a power supply and turn it on; you should get something like Figure 5-12.

Congratulations, you've successfully connected your Arduino and your iPhone via an XBee serial pass-through connection.

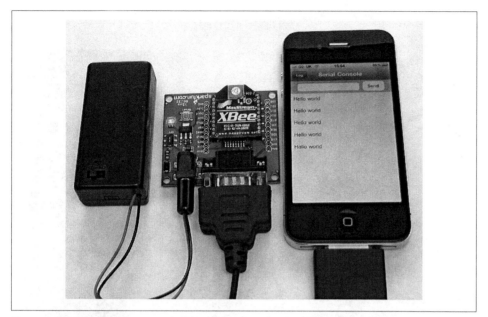

Figure 5-12. Everything connected and working

Going Further

A fairly trivial exercise at this point would be to take the sonar example from the previous chapter and substitute an XBee for the direct cable connection. There aren't any software changes needed, just hardware. Compare Figure 5-13 below to Figure 4-9 in the previous chapter.

If all goes well, you should see exactly the same thing you saw in the previous chapter.

Figure 5-13. Connecting the EZ1 example from Chapter 4 using XBee modules

CHAPTER 6

Other Ways to Connect

I've spent most of this book talking about the Redpark Serial Cable, and that's because it's the easiest way to connect external hardware directly to your iPhone or iPad. However, before the Redpark cable came along, we had to be somewhat more creative, and in this chapter, I'm going to look at some of the other ways you can connect external hardware to your iOS device.

Using the Network

I haven't talked about networking very much, but, as we're all well aware, the availability of ubiquitous data is the thing that changed everything. If we can Internet-enable our sensor, or other hardware, and make it part of the Internet of Things, our iPhone (or iPad) is already well equipped to talk to it.

Using Ethernet

The easiest way to Internet-enable a piece of arbitrary sensor hardware is (probably quite predictably by this stage) by making use of an Arduino. There are several ways you can accomplish this, but if you have a static sensor platform, the simplest way is probably to use wired Ethernet. See Figure 6-1.

 An Arduino shield is a board that can be plugged on top of the Arduino PCB, extending its capabilities. It may make use of some, or all, of the Arduino pins. However, those pins that are not used are generally passed through and exposed for other uses. See *http://shieldlist.org/* for a good list of the pin usage of many Arduino shields.

You can buy the official Arduino Ethernet Shield from various vendors, including both the Maker Shed (*http://www.makershed.com/ProductDetails.asp?ProductCode=MKSP7*) and SparkFun (*http://www.sparkfun.com/products/9026*).

Figure 6-1. Arduino Ethershield (top) and Arduino Uno (bottom)

Once you've attached the Ethernet shield to the Arduino, and plugged the shield into your network, a simple web client is fairly easy to put together. The code below establishes a network connection, and then carries out a Google search for the term "Arduino" and dispatches the page returned by the request to the serial connection:

```
#include <Ethernet.h>

byte mac[] = { 0xDE, 0xAD, 0xBE, 0xEF, 0xFE, 0xED };①
byte ip[] = { 144,173,229,19 };②
byte server[] = { 64, 233, 187, 99 }; // Google

Client client(server, 80);

void setup()
{
  Ethernet.begin(mac, ip);
  Serial.begin(9600);

  delay(1000);

  Serial.println("connecting...");

  if (client.connect()) {
    Serial.println("connected");
    client.println("GET /search?q=arduino HTTP/1.0");
    client.println();
  } else {
    Serial.println("connection failed");
```

```
    }
  }

  void loop()
  {
    if (client.available()) {
      char c = client.read();
      Serial.print(c);
    }

    if (!client.connected()) {
      Serial.println();
      Serial.println("disconnecting.");
      client.stop();
      for(;;)
        ;
    }
  }
```

❶ The Arduino Ethernet library requires that you define a MAC address for the Ethernet Shield. This is something that is normally done by the manufacturer, and most system administrators won't look kindly on your attempts to provide a random MAC for your Arduino. If you're going to connect your Arduino + Ethernet Shield to a network you do not directly control, you should consult with your system administrator beforehand.

❷ Another problem with the official Ethernet library is that it doesn't yet support DHCP and you must therefore supply your sketch with a static IP address. Again, this is something your system administrator might have issues with, as most modern networks make use of DHCP to provide dynamic allocation of IP addresses. If you're connecting your Arduino to a network you do not directly control, you should again consult your system administrator beforehand.

You can make use of the serial console provided by the Arduino development environment, or the SerialConsole application we wrote earlier in the book, to see the results returned by this sketch.

How do I generate a MAC address?

The purpose of a Medium Access Control (MAC) address is to distinguish physical devices on your network. Each device on your network must have a unique MAC address, and normally networked devices have a semiunique MAC embedded in their firmware by the manufacturer. However, in this case you are the manufacturer, and your sketch is (effectively) the firmware of the embedded device you are creating. Therefore, it's up to you to provide the MAC address for the device.

The first 6 octets of the MAC address are unique to the manufacturer of the device; the IEEE administers allocation of these addresses. However, you can assign a locally administered address in your code.

There are two bits in the first octet of a MAC address that are used to define certain aspects of the MAC address. The least significant bit of the first octet of the MAC address is the Individual/Group (I/G) address bit, while the next-to-least significant bit is the Universally or Locally (U/L) administered address bit.

The I/G bit is used to signify if the destination MAC address is a unicast or a multicast address. If the bit is set to 0, it is a unicast address. If the bit is set to 1, it is a multicast address. You should set this bit to 0 when assigning locally administered addresses.

The U/L bit is used to tell if the MAC address is the burned-in-address (BIA) or a locally administered address. When the bit is set to 1, the MAC address is recognized as a locally administered address.

There are therefore four sets of locally administered addresses that can be used on your network without fear of conflict, assuming there are no other developers making use of them. These are *X2:XX:XX:XX:XX:XX*, *X6:XX:XX:XX:XX:XX*, *XA:XX:XX:XX:XX:XX*, and *XE:XX:XX:XX:XX:XX* (where *X* should be replaced with an appropriate hexadecimal value).

A complete list of MAC address blocks assigned by the IEEE can be found at *http://standards.ieee.org/develop/regauth/oui/oui.txt*.

Perhaps more usefully, here is an example of how to build a simple web server that returns a short HTML document to the requesting client:

```
#include <Ethernet.h>

#define HTTP_HEADER "HTTP/1.0 200 OK\r\nServer: arduino\r\nContent-Type: text/html\r\n\r\n"

int LED = 9;

// network configuration.  gateway and subnet are optional.
byte mac[] = { 0xDE, 0xAD, 0xBE, 0xEF, 0xFE, 0xED };
byte ip[] = { 192, 168, 1, 99 };
byte gateway[] = { 192, 168, 1, 1 };
byte subnet[] = { 255, 255, 255, 0 };

Server server = Server(8080);

void setup()
{
  // sets the digital pin as output
  pinMode(LED, OUTPUT);

  // Start talking to the serial port
  Serial.begin(9600);
  Serial.println( "Start setup()");

  // initialize the ethernet device
  Serial.println("Ethernet.begin()");
  Ethernet.begin(mac, ip, gateway, subnet );
```

```
    // start listening for clients
    Serial.println("server.begin()");
    server.begin();

    // Flash the LED twice
    digitalWrite(LED, HIGH);
    delay(1000);
    digitalWrite(LED, LOW);
    delay(1000);
    digitalWrite(LED, HIGH);
    delay(1000);
    digitalWrite(LED, LOW);

    Serial.println( "Done setup()\n");

}

void loop(){
  Client client = server.available();
  if (client) {

    Serial.println("Connection from client");
    boolean current_line_is_blank = true;
    while (client.connected()) {
      digitalWrite(LED, HIGH);
      if (client.available()) {
        char c = client.read();
        Serial.print(c);

        if (c == '\n' && current_line_is_blank) {❶
          client.println( HTTP_HEADER );❷
          client.println();
          Serial.println("Sent header");

          client.println("<html><head><title>Test</title></head><body>");
          client.println("<p>Test String</p>");
          client.println("</body></html>");
          Serial.println("Sent page");
          break;
        }
        if (c == '\n') {
          current_line_is_blank = true;
        } else if (c != '\r') {
          current_line_is_blank = false;
        }
      }
    }
    client.stop();
    Serial.println("Closed client connection\n");
    digitalWrite(LED, LOW);

  }
}
```

❶ An HTTP request ends with a blank line. We look for it here so we can find the end of the request packet. If we've gotten to the end of the line (received a newline character) and the line is blank, the HTTP request has ended, and we can now send a reply.

❷ We send the standard HTTP HEADER response. This is not the same thing as the HTML <HEAD> block, but instead is a separate header that all web servers return before sending the requested HTML page.

While it's relatively simple to implement a web server, you may want to make use of the Webduino library instead. This is a generic web server library for the Arduino, and it can be found at *https://github.com/sirleech/Webduino*. It supports URL parameter parsing, handling of GET and POST methods, and web forms and images, as well as JSON and, importantly, support for RESTful interfaces out of the box.

The code below reimplements our simple web server using the generic library:

```
#include "SPI.h"
#include "Ethernet.h"
#include "WebServer.h"

static uint8_t mac[] = { 0xDE, 0xAD, 0xBE, 0xEF, 0xFE, 0xED };
static uint8_t ip[] = { 192, 168, 1, 99 };

#define PREFIX "/"
WebServer webserver(PREFIX, 80);

void helloCmd(WebServer &server, WebServer::ConnectionType type, char *, bool) {
  server.httpSuccess();
  if (type != WebServer::HEAD){
    P(helloMsg) = "<html><head><title>Test</title></head><body><p>Test String</
p><body></html>";
    server.printP(helloMsg);
  }
}

void setup() {
  Ethernet.begin(mac, ip);
  webserver.setDefaultCommand(&helloCmd);
  webserver.addCommand("index.html", &helloCmd);
  webserver.begin();
}

void loop() {
  char buff[64];
  int len = 64;
  webserver.processConnection(buff, &len);
}
```

Now that we have a web server, it's relatively simple to provide a RESTful interface returning a static piece of JSON with a current sensor reading embedded in it. Parsing such documents is the lifeblood of most iPhone applications and will be familiar to most of you. I'm not going to talk about it here, but I did cover it extensively in *Learning iPhone Programming (http://oreilly.com/catalog/9780596806439/)*.

Parsing JSON

JSON is a lightweight data-interchange format that is more or less human-readable but still easily machine-parsable. While XML is document-oriented, JSON is data-oriented. If you need to transmit a highly structured piece of data, you should probably render it in XML. However, if your data exchange needs are somewhat less demanding, JSON might be a good option.

The obvious advantage JSON has over XML is that since it is data-oriented and (almost) parsable as a hash map, there is no requirement for heavyweight parsing libraries. Additionally, JSON documents are much smaller than the equivalent XML documents. In bandwidth-limited situations, such as you might find on the iPhone, this can be important. JSON documents normally consume around half of the bandwidth of an equivalent XML document to transfer the same data.

While there is no native support for JSON in the Cocoa Touch framework, Stig Brautaset's *json-framework* library implements both a JSON parser and a generator and can be integrated into your project fairly simply. You can download the disk image with the latest version of the library from *http://code.google.com/p/json-framework/*.

Using WiFi

If your sensor bundle or equipment isn't stationary, or conveniently close to wired Ethernet, there are also WiFi shields available. One of the more commonly used shields is SparkFun's WiFly shield (see Figure 6-2). The WiFly shield can be bought directly from SparkFun (*http://www.sparkfun.com/products/9954*) for $89.95. A library to allow you to make easy use of the shield can be downloaded from *https://github.com/sparkfun/WiFly-Shield*. Unfortunately, the WiFly, along with most other WiFi shields currently on the market, is quite expensive.

Interestingly, having dealt with XBee radio modules earlier in the book, there is also now an XBee WiFi radio module that supports the IEEE 802.11 wireless standard rather than the IEEE 802.15.4 point-to-point standard. This module is pin-compatible with the existing Series 1 and 2 modules. More information can be found at *http://www.digi .com/products/wireless-wired-embedded-solutions/zigbee-rf-modules/point-multipoint -rfmodules/xbee-wi-fi#overview*.

If you're making extensive use of XBee modules, this option might allow you bridge your sensor network onto the Internet without a large additional expense.

Figure 6-2. Arduino WiFly Shield (top) and Arduino Diecimila (bottom)

Using a Soft Modem

Until the arrival of the Redpark cable, the easiest way to connect hardware directly was by using a soft modem. A well-known example of this practice is the Square credit card reader (see Figure 6-3).

Figure 6-3. The Square credit card reader

When you swipe the credit card with the Square reader, it reads the data and converts it into an analog audio signal. The audio signal is then picked up by the microphone, processed, and routed to the Square iOS app, where it is converted back into digital.

Switch Science Board

The Audio Jack Modem for iPhone and Android from Switch Science is available from SparkFun (*http://www.sparkfun.com/products/10331*) for $11.95 (see Figure 6-4).

Figure 6-4. The Switch Science SoftModem Board

This is a fairly simple board that more-or-less mimics the Square reader solution, except that it provides TTL serial. See *http://code.google.com/p/arms22/wiki/SoftModemBrea koutBoard* for more details about the SoftModem Board, along with sample code and the supporting libraries for both iOS and Arduino.

 The documentation for this board is in Japanese. However between Google Translate and the code samples available on the page, it's not too hard to work out how to use the board.

HiJack Board

While it makes use of similar technology, the HiJack board (see Figure 6-5) is slightly different from the Switch Science board. Interestingly, it includes its own microcontroller on board, which is powered by harvesting power directly from the iPhone. The harvester can supply 7.4 mW to a load with 47% power conversion efficiency when driven by a 22 kHz tone. It was developed at the University of Michigan and was intended to allow the development of simple medical hardware that could be easily deployed into the developing world. More information about HiJack project can be found at *http://www.eecs.umich.edu/~prabal/projects/hijack/*.

Figure 6-5. The HiJack board (right) and programmer (left)

You can purchase a HiJack board as part of a development kit that includes a board, a programmer, two proto-boards, and jumper wires from Seeed Studio (*http://www .seeedstudio.com/depot/hijack-development-pack-p-865.html*) for $79.

More information about the development kit can be found on the vendor's wiki site (*http://garden.seeedstudio.com/index.php?title=Hijack#Introduction*).

Using the MIDI Protocol

With the arrival of the iPad and the camera connection kit (available from the Apple Store, *http://store.apple.com/us/product/MC531ZM/A*), it became possible to directly connect USB devices to the iPad (see Figure 6-6). While most USB-HID profiles are not supported, and the connector was originally intended only to allow you to transfer images from a connected camera onto your iPad, there is an interesting exception. The connector also allows your iPad to talk directly to USB MIDI devices.

 Before iOS 4.x, the USB dongle from the iPad Camera Connection Kit also allowed you to connect USB accessories to the iPhone 4. Unfortunately, with this revision of the operating system, a change to the phone's firmware was made that meant that a lower amperage was provided to the Dock connector and USB devices no longer had enough power to function.

There are a number of MIDI shields for the Arduino on the market including the SparkFun shield (*http://www.sparkfun.com/products/9595*) (see Figure 6-7). Figure 6-8 shows this shield connected via a Screw Shield to an Arduino Uno board. This project (see *https://github.com/bjepson/iPad-MIDI-Simple-Demo* for more details) al-

Figure 6-6. The USB dongle from the iPad Camera Connection Kit

Figure 6-7. Arduino Uno with MIDI shield

lows you to control off-the-shelf MIDI software on your iPad directly from an Arduino board.

Interestingly, the availability and flexibility of the MIDI protocol means that when talking to an Arduino, pretending to be a MIDI device, there isn't any real need for the MIDI protocol to do what most people expect of it. Effectively we can pervert the protocol to allow you to embed arbitrary serial data in the MIDI packets. While I doubt

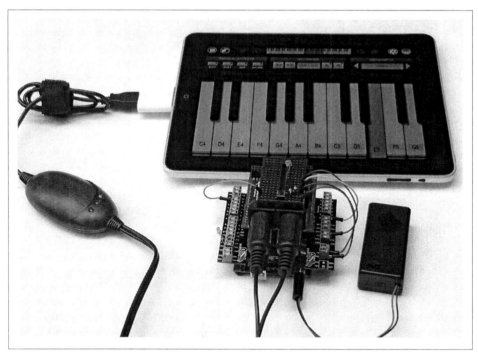

Figure 6-8. An Arduino connected to an iPad running standard off-the-shelf MIDI-enabled software via a MIDI shield and the Camera Connection Kit

such an application abusing the protocol would make it past the review team into the App Store (by some measures of things quite rightly so), it is possible.

The HIDDUINO

The latest generation of Arduino boards, including the Arduino Uno, replaces the commonly used FTDI USB chip with an Atmega8u2 microcontroller. This is an interesting development, as this means that the latest generation of Arduinos are effectively dual-core, and if you're not intending to make use of the USB connection, the secondary microcontroller intended for USB serial communication can be reused for other purposes. Among these would be to replace the current firmware with new firmware conforming to the USB-HID protocol. In essence, this means the Arduino board can appear to be a mouse, keyboard, joystick, or any other USB device to the host computer. Importantly, this can be done without the addition of any new driver software on the host computer end. To your Mac, the Arduino will just look like a normal USB device.

The Atmega8u2 can be easy reprogrammed with an off-the-shelf AVR programmer or flashed with a new firmware using the DFU bootloader.

One of the things that can be done is to reflash the microcontroller to enable the Arduino to be seen as a MIDI HID device, thus removing the need for custom interpretation or conversion software. MIDI-enabled programs on your iPad will be able to see the Arduino directly, without the need for a MIDI shield or other hardware. You'll also be able to talk to the Arduino board using the MIDI framework, which is part of the iOS SDK. See *http://mtiid.calarts.edu/research/hiduino* for more details.

Summary

The method you use to connect hardware to your iOS device obviously depends quite critically on what hardware you're using, and why you wanted to connect the hardware to your iPhone or iPad in the first place. Hopefully this chapter has pointed you toward some alternative solutions if directly connecting the hardware to your iPhone via a serial connection isn't what you needed to do.

About the Author

Alasdair Allan is the author of *Learning iPhone Programming* (*http://oreilly.com/cata log/9780596806439/*), *iOS Sensor Programming* (*http://oreilly.com/catalog/ 9781449390334/*), *Basic Sensors in iOS* (*http://oreilly.com/catalog/9781449308469/*), *Geolocation iOS* (*http://oreilly.com/catalog/9781449308445/*), *iOS Sensor Apps with Arduino* (*http://oreilly.com/catalog/9781449308483/*), and *Augmented Reality in iOS* (*http://oreilly.com/catalog/9781449308506/*), all published by O'Reilly Media. He is a senior research fellow in astronomy at the University of Exeter. As part of his work there he is building a distributed peer-to-peer network of telescopes which, acting autonomously, will reactively schedule observations of time-critical events. Notable successes include contributing to the detection of the most distant object yet discovered, a gamma-ray burster at a redshift of 8.2. Alasdair also runs a small technology consulting business (*http://www.babilim.co.uk/*) writing bespoke software, building open hardware, and providing training. He sporadically writes blog posts (*http://www.dai lyack.com/*) about things that interest him, or more frequently provides commentary about them in 140 characters or less (*http://twitter.com/#!/aallan/*).

Get even more for your money.

Join the O'Reilly Community, and register the O'Reilly books you own. It's free, and you'll get:

- $4.99 ebook upgrade offer
- 40% upgrade offer on O'Reilly print books
- Membership discounts on books and events
- Free lifetime updates to ebooks and videos
- Multiple ebook formats, DRM FREE
- Participation in the O'Reilly community
- Newsletters
- Account management
- 100% Satisfaction Guarantee

Signing up is easy:

1. Go to: oreilly.com/go/register
2. Create an O'Reilly login.
3. Provide your address.
4. Register your books.

Note: English-language books only

To order books online:

oreilly.com/store

For questions about products or an order:

orders@oreilly.com

To sign up to get topic-specific email announcements and/or news about upcoming books, conferences, special offers, and new technologies:

elists@oreilly.com

For technical questions about book content:

booktech@oreilly.com

To submit new book proposals to our editors:

proposals@oreilly.com

O'Reilly books are available in multiple DRM-free ebook formats. For more information:

oreilly.com/ebooks

O'REILLY®

Spreading the knowledge of innovators

oreilly.com

The information you need, when and where you need it.

With Safari Books Online, you can:

Access the contents of thousands of technology and business books

- Quickly search over 7000 books and certification guides
- Download whole books or chapters in PDF format, at no extra cost, to print or read on the go
- Copy and paste code
- Save up to 35% on O'Reilly print books
- **New!** Access mobile-friendly books directly from cell phones and mobile devices

Stay up-to-date on emerging topics before the books are published

- Get on-demand access to evolving manuscripts.
- Interact directly with authors of upcoming books

Explore thousands of hours of video on technology and design topics

- Learn from expert video tutorials
- Watch and replay recorded conference sessions

Spreading the knowledge of innovators   safari.oreilly.com

Lightning Source UK Ltd.
Milton Keynes UK
UKOW021038220413

209568UK00004B/214/P